Smart-Kids

Carol Every, Michelle Faure

Mathematics

奔跑吧数学

英汉对照

2级
Grade 2

探索奇妙的数学世界

〔英〕卡罗尔·艾瑞
　　　米歇尔·福尔　主编

赵媛媛　译

天津出版传媒集团

天津科技翻译出版有限公司

著作权合同登记号：图字：02-2016-11

图书在版编目(CIP)数据

奔跑吧,数学：探索奇妙的数学世界. 2 级：英汉对照/(英)卡罗尔·艾瑞(Carol Every),(英)米歇尔·福尔(Michelle Faure)主编；赵媛媛译. —天津：天津科技翻译出版有限公司,2016.8

书名原文：Smart-Kids Mathematics：Grade 1

ISBN 978-7-5433-3624-7

Ⅰ. ①奔⋯　Ⅱ. ①卡⋯ ②米⋯ ③赵⋯　Ⅲ. ①数学-儿童读物-英、汉　Ⅳ. ① O1-49

中国版本图书馆 CIP 数据核字(2016)第 158066 号

Authorized reprint from the English language edition, entitled Smart-Kids Mathematics: Grade 1, ISBN 978-1-77025-727-6 by Carol Every, Michelle Faure, published by Pearson Education, Inc., publishing as Pearson Education South Africa (Pty) Ltd, Copyright © 2010.

All rights reserved. No part of this book may be reproduced or transmitted in any form or by any means, electronic or mechanical, including photocopying, recording or by any information storage retrieval system, without permission from Pearson Education Inc.

English Reprint/adaptation published by Pearson Education Asia Limited Copyright © 2016.

中文简体字版权属天津科技翻译出版有限公司。

出　　　版	天津科技翻译出版有限公司
出 版 人	刘　庆
地　　　址	天津市南开区白堤路 244 号
邮政编码	300192
电　　　话	(022)87894896
传　　　真	(022)87895650
网　　　址	www.tsttpc.com
印　　　刷	天津市银博印刷集团有限公司
发　　　行	全国新华书店

版本记录：880×1230　16 开本　5.25 印张　100 千字
　　　　　2016 年 8 月第 1 版　2016 年 8 月第 1 次印刷
　　　　　定价：29.80 元

(如发现印装问题,可与出版社调换)

出版者的话

《奔跑吧，数学：探索奇妙的数学世界》(英汉双语)(1~4级)是从国际著名教育出版机构英国培生教育集团引进的数学学习益智书，真实反映了国外小学生的现行教学内容，全面展现了国外小学生丰富多彩的学习场景。

为什么很多小学生不喜欢学习数学，学习效果不好，没有学习的兴趣？这恐怕和我们侧重于背公式，做习题，准备考试，这种比较枯燥的学习方式不无关系。这套丛书全面体现国外小学生要掌握的数学基础知识和英语表达，展现生动活泼的学习和游戏场景。读者可从中领会原汁原味的国外小学生的学习内容，学习简单的英语表达。同时，书中着重通过游戏让孩子亲自动手，寓教于乐，图文并茂，让孩子在提高动手能力的同时提高学习数学的兴趣。通过游戏的方式，让孩子在奇妙的数学世界中快乐地奔跑、遨游和探索。

书后备有小贴纸，可以增加孩子学习的乐趣。并贴心地配有"注释"，介绍了每个小游戏的训练目的和训练方法，帮助孩子和家长一起打开学习数学的大门。每册学习完成后，家长可以为孩子颁发"证书"，让孩子拥有满满的成就感。

这套丛书采用英汉双语对照的形式，既保留了原版英文，介绍了原汁原味的英语背景和地道的英语口语表达，又为方便孩子理解可以独立完成练习而增加了中文翻译，在学习数学技能、提高数学学习能力的同时，也能提高英语水平，可谓一书在手，一举两得。

目录
CONTENTS

Term 1
- Snake patterns 小蛇花纹 2
- Boxes and balls 盒子和球 3
- Leapfrog 跳跃的青蛙 4
- Sort it out 分类 5
- My day 我的一天 6
- Counting teddies 数泰迪熊 7
- Heavy or light? 重还是轻? 8
- Shape up 补充完整 9
- Jody's birthday Jody 的生日 10
- Yummy treats! 美味佳肴! 11
- Can you guess? 你能猜到吗? 12
- Busy bees 忙碌的蜜蜂 13
- Longer or shorter time? 哪个花费时间更长，哪个更短? 14
- Rock hops 跳房子游戏 15
- Put your foot on it! 用脚测量! 16
- What is the same? 哪些属性相同? 17
- Yesterday, today, tomorrow 昨天，今天，明天 18
- Button up 数纽扣 19
- Can you see me? 你能看见我吗? 20

Term 2
- Decorating cakes 装饰蛋糕 21
- Many marbles 很多弹珠 22
- Sharing sweets 分享糖果 23
- Fun with food 喜欢的食物 24
- Counting toys 数玩具 25
- All shapes and sizes 大大小小各种形状 26
- Bunny hops 兔子跳 27
- Double up 双倍 28
- How long does it take? 需要多长时间? 29
- Treasure hunt 寻宝 30
- Serious sewing 复杂的缝纫 31
- Munching leaves 咀嚼树叶 32
- Spoon it in! 用勺子舀水! 33
- Sports day 运动会 34
- Lots of spots 很多的斑点 35
- Money fun 存钱的乐趣 36
- Emma's new crayons Emma 的新蜡笔 37
- Royal mats 城堡毯子 38
- Clever guesses 聪明的猜测 39

Term 3
- Number gym 数字游戏 40
- Shapes and things 形状和物品 41
- All aboard 请上车 42
- Buckets and bowls 桶和碗 43
- Same, same 相同，相同 44
- Show of hands 伸出双手 45
- I spy 我来观察 46
- How many marbles? 有多少颗弹珠? 47
- Doubling with dominoes 骨牌加倍 48
- Lebo's beady eye Lebo 雪亮的眼睛 49
- At school 在学校 50
- Toy box tally 给玩具箱做标记 51
- Let's go shopping 让我们去购物 52
- Winter woollies 冬天的毛线物品 53
- Missing numbers 缺少的数字 54
- Tropical birds 热带鸟儿 55
- Computer fun 计算机游戏 56
- Hands on 用手测量 57

Term 4
- Peek-a-boo! 躲猫猫! 58
- Body snap! 身体快照! 59
- Jody the witch 女巫 Jody 60
- Weigh to go! 称重! 61
- Where is it? 在哪里? 62
- Sorted! 分类! 63
- Poles and ladders 杆子和梯子 64
- Fancy flowers 美丽的鲜花 66
- Keep counting 继续数数 67
- High five 伸手击掌 68
- What is the pattern? 这是什么规律? 69
- Happy birthday! 生日快乐! 70
- Happy birthday, Gran! 奶奶，生日快乐! 71
- Hide and seek 捉迷藏 72
- Shop till you drop 逛逛商店 74
- Does it belong? 物品归类? 75
- Notes 注释 76

奔跑吧，数学
探索奇妙的数学世界

Smart-Kids
Mathematics Grade 2

英汉对照
2级

Draw a picture of yourself with two friends.

画一张自己和两个朋友的画像。

Write how old you are.

写下你的年龄。

_____

Snake patterns
小蛇花纹

The snakes have lost some of their patterns.
小蛇身上丢了一些花纹。

1. Complete the patterns. 请将花纹补充完整。

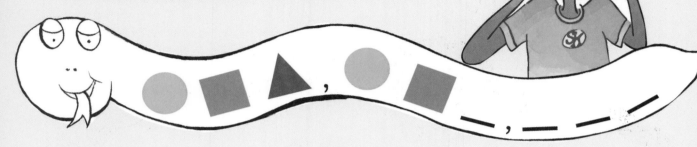

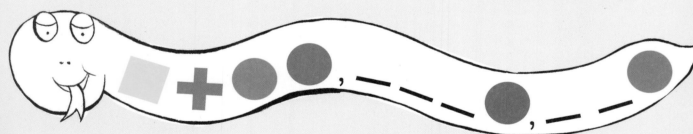

2. Complete the number pattern. 请把数字花纹补充完整。

Boxes and balls 盒子和球

1. Draw a **square** around the objects that look like boxes.
   在像盒子形狀的物品周围画一个正方形。
2. Draw a **circle** around the objects that look like balls.
   在像球形状的物品周围画一个圆形。

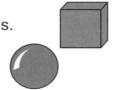

Leapfrog 跳跃的青蛙

The frogs are racing to the finishing line.
Which one will get there first?
Which one will be second?
Which one will come third?

几只青蛙正在比赛,看谁先跳完数字线。谁会是第 1 名？谁会是第 2 名？谁会是第 3 名？

1. How many more leaps for each frog to get to 5? 每只青蛙要跳几下才能跳到 5？
   Write your answer in each block. 在题后的方块中写出答案。

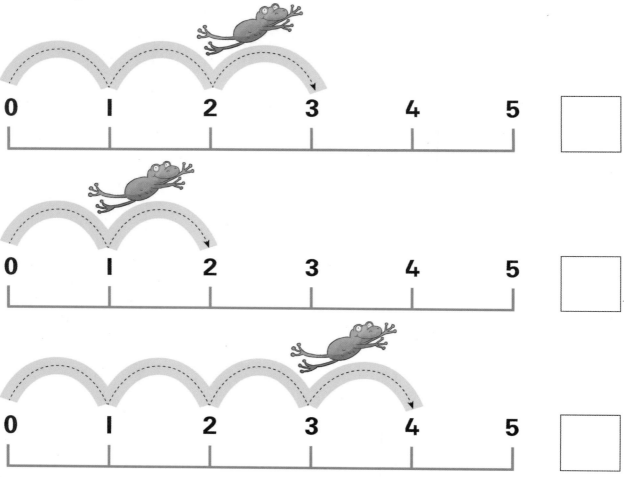

2. Which numbers on the number lines are smaller than 2?
   在上述数字线中哪些数字小于 2？

3. Which numbers on the number lines are bigger than 3?
   在上述数字线中哪些数字大于 3？

4. Use one of the number lines to help you fill in the answers.
   用上述一条数字线帮你写出答案。

   3 + ☐ = 5   5 − ☐ = 2

Sort it out
分类

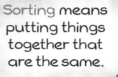

Sorting means putting things together that are the same.

分类就是把相同的物品放在一起。

Do you know how to sort objects?
你知道怎么分类吗?

1. Collect the cutlery in your kitchen drawers – except for the sharp knives! 收集厨房抽屉里的餐具——不包括锋利的刀!
2. Put all the spoons together. Put all the forks together. Put all the knives together.
   把所有的勺子放在一起。把所有的叉子放在一起。把所有的刀子放在一起。

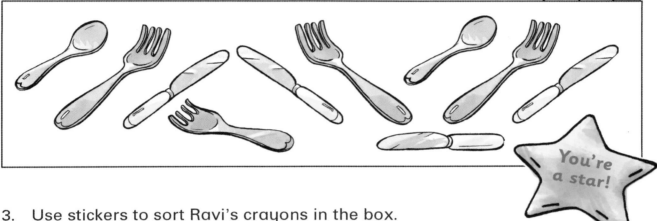

3. Use stickers to sort Ravi's crayons in the box.
   用贴纸将 Ravi 的蜡笔分类放在盒子里。

My crayon box 我的蜡笔盒

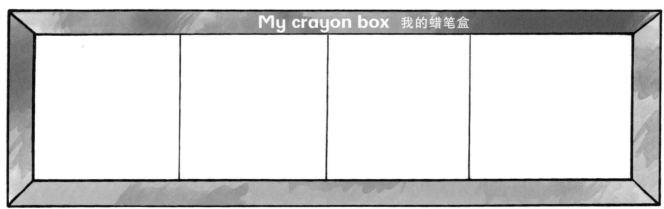

4. Of which colour crayon does Ravi have the most?
   Ravi 哪个颜色的蜡笔最多?

My day 我的一天

Lebo writes about what she does each day in her diary.
Lebo 在日记中写出自己每天的活动。

1. Help her finish the sentences. Choose from | morning | night | afternoon |
   帮她完成下列句子。从"morning""night""afternoon"中选择词语。

In the _____,
I wake up early. I put on my clothes, eat my breakfast and go to school.

在___，我早早地起床。我穿好衣服，吃早饭，然后去上学。

At _____,
I eat my supper. I have a bath. I brush my teeth and go to bed.

在___，我先吃晚饭，然后洗澡，再刷牙，最后去睡觉。

In the _____,
I do my homework. I play with my toys, eat my lunch and watch television.

在___，我先做作业，然后玩会儿玩具，再吃午饭并看会儿电视。

2. What happens first?
   What happens second?
   What happens third?
   Write 1, 2 or 3 in the blocks in the pictures.

   哪种情况最先发生？
   哪种情况其次发生？
   哪种情况最后发生？将1,2,3写在图片的方块中。

Counting teddies
数泰迪熊

1. Emma has a collection of teddies. Count them.
   Emma 收藏了很多泰迪熊。数一下吧。

   Emma has ☐ teddies altogether.

   Emma 一共有____只泰迪熊。

   Emma's granny gives her three more teddies. Draw them.
   Emma 的奶奶又给了她 3 只泰迪熊。把它们画出来。
   How many teddies does she have now?
   Circle your answer.
   她现在有多少只泰迪熊？圈出你的答案。

   6   8   9

2. Lebo also has a collection of teddies. Count them.
   Lebo 也收藏了很多泰迪熊。数一下。

   Lebo has **6** teddies altogether.

   Lebo 一共有 6 只泰迪熊。

   Lebo's grandpa gives her four more teddies. Draw them.
   Lebo 的爷爷又给了她 4 只泰迪熊。把它们画出来。
   How many teddies does she have now?
   Circle your answer.
   她现在有多少只泰迪熊？圈出你的答案。

   7   10   8

3. Who has the most teddies? Tick one name.
   谁的泰迪熊比较多？在她的名字旁打"√"。

   Lebo ☐   Emma ☐

Heavy or light? 重还是轻?

How heavy do you think you are?
你认为自己有多重？

1. Draw a picture of yourself in the frame.
   在中间的框中画一幅自画像。

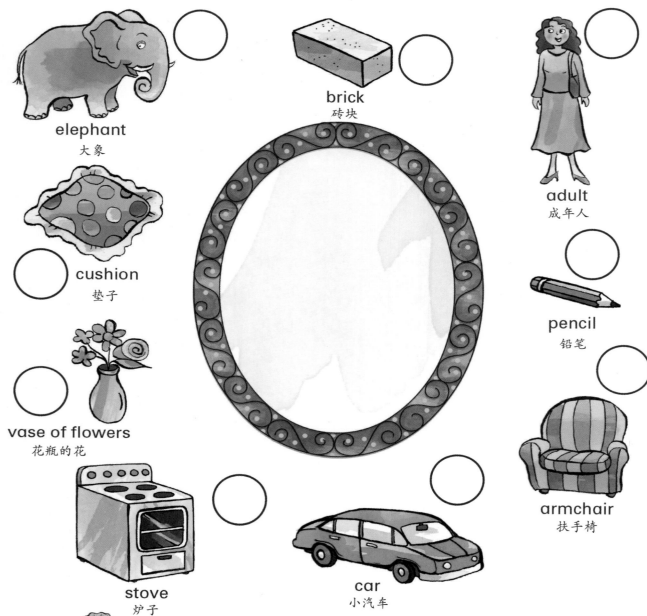

2. Put a sticker next to the things that are lighter than you.
   将一张"l"贴纸粘在比你轻的物品旁边。

   Put a [h] sticker next the to things that are heavier than you.
   将一张"h"贴纸粘在比你重的物品旁边。

3. Draw a circle around the heaviest object.
   在最重的物品上画一个圆圈。

4. Draw a square around the lightest object.
   在最轻的物品上画一个方框。

Shape up 补充完整

1. Find and complete the patterns.
   发现规律，并按规律补充完整。

1 1, 2 _, 3 _, 4 _, 5 _, 6 6, 7 7

10, 9, 8, 7, 6, 5, _, _, _, 1

2. Make your own patterns.
   自己做出规律。

   a. Use shapes here: 用图形做规律：

   b. Use numbers here: 用数字做规律：

Jody's birthday

Jody 的生日

Jody is very excited because soon it will be her birthday. Look at the calendar.

Jody 非常高兴，因为她马上要过生日了。看一下日历。

1. Jody's birthday is on 11 August. Colour that block yellow. On what day of the week is Jody's birthday?

   Jody 的生日是 8 月 11 日。将这一天的方框涂成黄色。Jody 的生日是本周的星期几？

   _____

2. Jody's party is on 13 August. Colour that block green. On what day of the week is her party?

   Jody 的聚会是 8 月 13 日。将这一天的方框涂成绿色。Jody 的聚会是在本周的星期几？

   _____

August						
Sunday	Monday	Tuesday	Wednesday	Thursday	Friday	Saturday
	1	2	3	4	5	6
7	8	9	10	11	12	13
14	15	16	17	18	19	20
21	22	23	24	25	26	27
28	29	30	31			

3. Jody is sending out her party invitations on 2 August.

   Jody 在 8 月 2 日发出她的聚会邀请函。

   Colour that block red. On what day of the week will she send them?

   将这一天涂成红色。她是本周星期几发的邀请函？

   _____

4. Colour the first day of the month blue.

   将本月的第一天涂成蓝色。

5. Colour the last day of the month orange.

   将本月的最后一天涂成橙色。

Yummy treats!

美味佳肴！

Lebo's mum is baking. Help her to share the teatime treats. Lebo 的妈妈在做烘焙。帮她分一下这些美食。

1. Mum bakes 4 gingerbread men. She shares them equally between Jaco and Lebo.
   妈妈烤了 4 块美味的小人饼干。她要把这 4 块饼干平均分给 Jaco 和 Lebo。

Work it out. 写出算式。

How many do they each get?
他们每个人可以分几块？

2. Mum also bakes 5 cupcakes. She shares them equally between the children. Are there any left?
   妈妈还烤了 5 块纸杯蛋糕。她也要平均分给两个孩子。会有剩余的吗？

Work it out. 写出算式。

How many does each child get?
每个孩子可以分到几块？

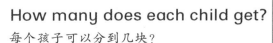

Can you guess?

你能猜到吗？

Are you good at guessing?
你擅长猜测吗？

> An estimate is a careful guess before you count.
>
> 在你计数之前，估计是一种小心的猜测。

1. Look quickly at each picture.
   Guess how many dots there are.
   快速地看一遍每张图。猜猜上面共有几个点。

2. Write down your estimate in each green box.
   在每个绿框中写下你猜测的数字。

3. Now count the dots in each picture.
   Write down the correct number in each red box.
   Write down the word for the number you counted on the line.
   现在数一下每张图中的点数。在每个红框中写出正确的点数。在横线上写出与你数出的数量对应的单词。

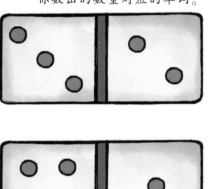

estimate
估计数

count
实际数

estimate
估计数

count
实际数

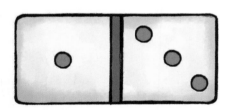

estimate
估计数

count
实际数

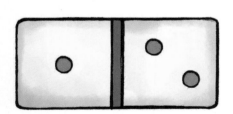

estimate
估计数

count
实际数

4. Which of the numbers are smaller than 5?
   Colour the orange blocks blue.
   哪些图片点数小于5？将橙色图片涂成蓝色。

Busy bees
忙碌的蜜蜂

These bees have lost their way home.
Can you help them get to the correct hive?
这些蜜蜂找不到回家的路了。你能帮它们找到正确的蜂巢吗?

1. Draw a line to match each bee to the correct hive. 将每个蜜蜂与正确的蜂巢之间进行连线。
Use the number line to help you. 你可以借助数字线。

2. Which hive has the biggest number? Draw a circle around it.
哪个蜂巢号码数最大？用圆圈画出来。

Which hive has the smallest number? Draw a block around it.
哪个蜂巢号码数最小？用方框画出来。

Longer or shorter time?

哪个花费时间更长,哪个更短?

Some things take a longer time. Some things take a shorter time.
下列图中一些事情花费更长的时间。一些事情花费更短的时间。

Draw a circle around the thing in each box that takes a shorter time.
在下列每个图框中,用圆圈标出花费时间更短的事情。

washing your hands 洗手   bathing 洗澡

eating supper 吃晚饭   eating a biscuit 吃饼干

 Mandla

drawing a picture 画一张画   writing your name 写名字

a flower to bloom 一朵花开   a tree to grow 一棵树长大

Rock hops 跳房子游戏

Emma steps on every rock as she hops across and back. Help her to cross the river without getting her feet wet!

Emma 踩在每块岩石上跳来跳去。帮助她过河而又不弄湿脚。

1. Start at 0. Count the rocks as Emma hops on them.
   从 0 号岩石开始，Emma 数着数一个个地跳。
2. Start at 20. Count backwards to 0.
   从 20 号岩石开始，跳回 0 号岩石。
3. Which numbers are smaller than 10? Colour those rocks yellow.
   哪些数字比 10 小？将那些岩石涂成黄色。
4. Which numbers are bigger than 10? Colour those rocks blue.
   哪些数字比 10 大？将那些岩石涂成蓝色。

Put your foot on it!

用脚测量！

You can measure with your feet.
你可以用脚测量。

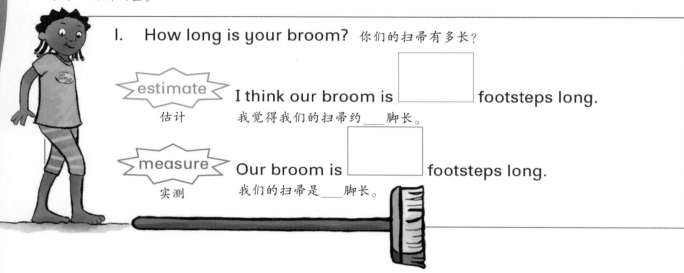

1. How long is your broom? 你们的扫帚有多长？

   estimate 估计 — I think our broom is ☐ footsteps long.
   我觉得我们的扫帚约___脚长。

   measure 实测 — Our broom is ☐ footsteps long.
   我们的扫帚是___脚长。

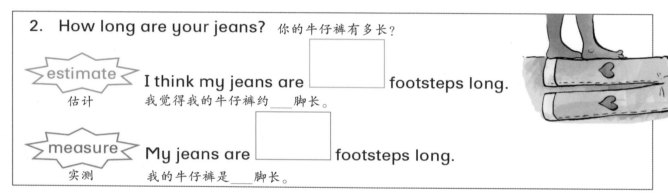

2. How long are your jeans? 你的牛仔裤有多长？

   estimate 估计 — I think my jeans are ☐ footsteps long.
   我觉得我的牛仔裤约___脚长。

   measure 实测 — My jeans are ☐ footsteps long.
   我的牛仔裤是___脚长。

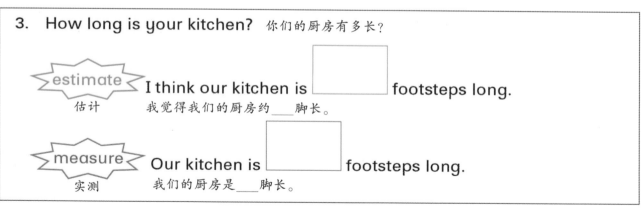

3. How long is your kitchen? 你们的厨房有多长？

   estimate 估计 — I think our kitchen is ☐ footsteps long.
   我觉得我们的厨房约___脚长。

   measure 实测 — Our kitchen is ☐ footsteps long.
   我们的厨房是___脚长。

4. Which is the shortest? _____
   上述哪个的长度最短？

5. Which is the longest? _____
   上述哪个的长度最长？

What is the same? 哪些属性相同？

Things that belong together have something about them that is the same.
放在一起的物品有一些相同之处。

1. Look at what Jody collected. 看一看 Jody 收集的物品。
   What is the same about them? _____
   它们的相同之处是什么？

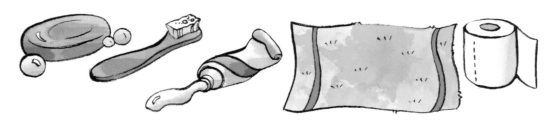

2. Look at what Emma collected. 看一看 Emma 收集的物品。
   What is the same about them? _____
   它们的相同之处是什么？

3. Look at what Jaco collected. 看一看 Jaco 收集的物品。
   What is the same about them? _____
   它们的相同之处是什么？

4. Collect 5 things that belong together. Draw them.
   收集5种有相同之处的物品。并把它们画出来。

5. What is the same about them? _____
   它们的相同之处是什么？

Term 1

Skill: collecting and recording data

17

Yesterday, today, tomorrow

昨天,今天,明天

Ravi is on holiday at a game reserve. Help him write a letter to his granny at home.

Ravi 在一个野生动物保护区度假。帮他给在家的奶奶写封信。

Use these words to complete the sentences:

用"Tomorrow""Today""Yesterday"来完成句子。

| Tomorrow | Today | Yesterday |

6 April
4月6日

Dear Gran
亲爱的奶奶

5 April

_____ we went to see the lions.
_____ 我们去看了狮子。

6 April

_____ I am at the waterhole. I can see a hippo.
_____ 我来到水池边,看到一只河马。

7 April

_____ we are going to see the elephants.
_____ 我们打算去看大象。

Love from
Ravi
爱你的 Ravi

18

Button up 数纽扣

Emma is choosing pretty buttons for her new dress.
Emma 在为她的新裙子挑选一些漂亮的纽扣。

1. First Emma chooses 2 buttons. Then she chooses another 3 buttons.
   How many buttons does Emma have altogether?
   Draw the buttons in the number sentence below.
   Emma 先选了 2 粒纽扣，然后她又选了另外 3 粒。Emma 一共挑选了多少粒纽扣？将这些纽扣画在下面的算式中。

    +  =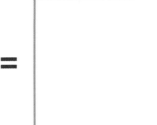

2. Jody puts 6 buttons on the table. She gives 2 to her mother.
   How many does she have left? Draw the buttons in the number sentence below.
   Jody 将 6 粒纽扣放在桌子上。然后她把其中 2 粒给了她妈妈。Jody 还有几粒纽扣？将这些纽扣画在算式中。

    −  =

3. Who has the fewest buttons? Tick one name.
   谁的纽扣最少？在名字旁画"√"。

   Jody ☐   Emma ☐

4. Count these buttons: 数一下这些纽扣：

   ☐ buttons

   共＿＿粒纽扣

Can you see me? 你能看见我吗?

We can play hide-and-seek.
我们来玩捉迷藏的游戏。

1. Colour the thing that is: 按要求将下列物品涂色:

   behind the box 盒子后面的物品　　　　in front of the box 盒子前面的物品

2. Draw: 按要求画出来:
   - a hat in front of the box　在盒子前面画一顶帽子
   - a shoe next to the box　在盒子旁边画一只鞋子
   - a ball on top of the box　在盒子顶部画一个球
   - a table under the box　在盒子下面画一张桌子

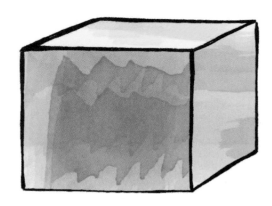

Decorating cakes
装饰蛋糕

Emma and Jaco love baking.
Let's see what they do.

Emma 和 Jaco 都热爱烘焙。我们来看看他们做了什么。

Emma has **7** cherries to put on her cake. She eats **2** of them.

Emma 有 7 颗樱桃可以放在蛋糕上,她吃了其中 2 颗。

1. How many does she have left?
   她还有几颗樱桃?

2. Write the number sentence below.
   将算式写在下面。

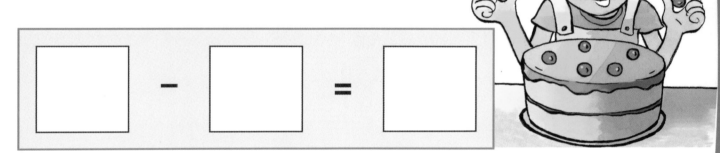

Jaco puts **4** cherries on his cake. Then he adds another **4**.

Jaco 把 4 颗樱桃放在蛋糕上。然后他又放上另外 4 颗。

3. How many cherries does his cake have?
   他的蛋糕上一共有几颗樱桃?

4. Write the number sentence below.
   将算式写在下面。

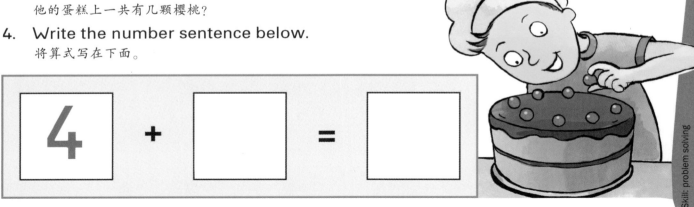

Many marbles

很多弹珠

Help the children count their marbles. Use the number line to help you.

帮孩子们数一下他们的弹珠。可以借助下面的数字线帮忙计算。

1. Lebo has 4 marbles. Emma gives her 3 more. And another 3.
   Lebo 有 4 个弹珠。Emma 又给了她 3 个。然后又给了她 3 个。

    + 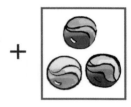 +    Now Lebo has ☐ marbles.
   现在 Lebo 有___个弹珠。

2. Jody has 4 marbles. Mandla gives her 2 more. And another 2.
   Jody 有 4 个弹珠。Mandla 又给了她 2 个。然后又给了她 2 个。

    + 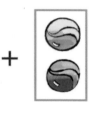 +    Now Jody has ☐ marbles.
   现在 Jody 有___个弹珠。

3. Jaco has 6 marbles. Ravi gives him 2 more. And another 2.
   Jaco 有 6 个弹珠。Ravi 又给了他 2 个,然后又给了他 2 个。

   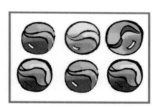 + +    Now Jaco has ☐ marbles.
   现在 Jaco 有___个弹珠。

4. Mandla has 2 marbles. Lebo gives him 4 more. And another 4.
   Mandla 有 2 个弹珠。Lebo 又给了他 4 个,然后又给了他 4 个。

   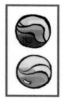 +  + 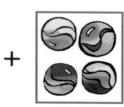   Now Mandla has ☐ marbles.
   现在 Mandla 有___个弹珠。

Sharing sweets
分享糖果

It is always fun to share sweets with a friend. Share the sweets equally between Emma and Lebo. Write the answers.

和朋友分享糖果是一件很开心的事。将糖果平均分给 Emma 和 Lebo。写下答案。

1. Half of 10 is ☐. 10颗糖果一人一半是___。

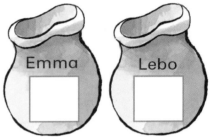

2. Half of 8 is ☐. 8颗糖果一人一半是___。

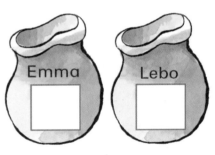

3. Half of 6 is ☐. 6颗糖果一人一半是___。

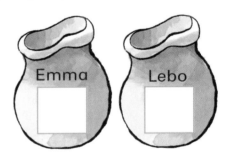

4. Half of 4 is ☐. 4颗糖果一人一半是___。

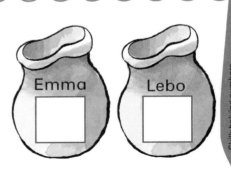

Fun with food

喜欢的食物

What is your favourite food? What do you like to drink?

你最喜欢什么食物？你喜欢喝什么？

Use your stickers to find the number names that match the pictures.

用书后的贴纸标出与下图相匹配的食物数量。

Counting toys

数玩具

Mandla is helping Emma sort out her toy box. Can you help the children?

Mandla 在帮助 Emma 将玩具分类。你能帮帮他们吗?

0　1　2　3　4　5　6　7　8　9　10

1. Count all the toys. Then write the answers. 数一下一共有多少个玩具,然后写出答案。
   Write the number and its name. 写出玩具的数量和名称。

2. Tick an answer. 在答案后面画"√"。
   Most of the toys are:　balls ☐　kites ☐　dolls ☐ .
   数量最多的玩具是：　　球　　　风筝　　　娃娃

3. The smallest number of toys is _____ . 数量最少的玩具是___个。

4. Choose the correct word from the box. 从"more""fewer"中选择正确的词完成句子。

   There are ⬜ more ⬛ fewer than 5 of every toy. 每种玩具都比 5 个 多或少。

All shapes and sizes

大大小小各种形状

There are many shapes all around us.
You just have to look for them!
我们身边有很多形状。你现在就去寻找它们。

1. Which shapes have round sides? Colour them red. Match them to the objects.
   哪些形状是圆形的边？将它们涂成红色。将它们与实际物体相匹配。

2. Which shapes have straight sides? Colour them yellow. Match them to the objects. 哪些形状是直线形的边？将它们涂成黄色。将它们与实际物体相匹配。

3. Match the small shapes to the small objects. Join them with lines.
   将小的形状与小的物体相匹配，并做连线。

4. Match the big shapes to the big objects. Join them with lines.
   将大的形状与大的物体相匹配，并做连线。

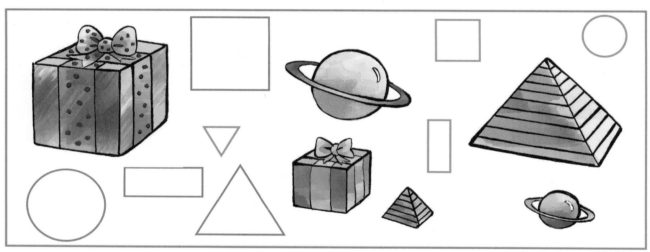

5. Draw the same shapes, but smaller. 画出形状相同但更小一些的物体。

6. Draw the same shapes, but bigger. 画出形状相同但更大一些的物体。

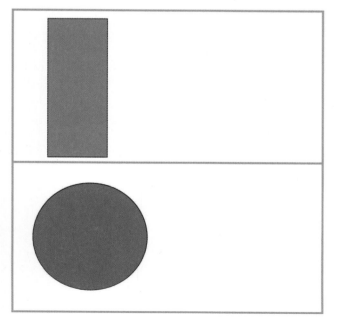

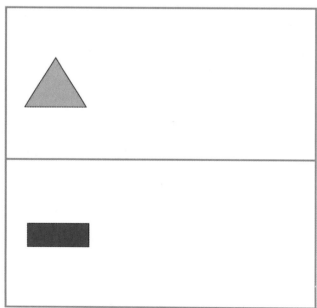

Bunny hops
兔子跳

**Look at the rabbit hopping to get his carrot.
What do you notice?**

看看小兔子正蹦跳着去拿胡萝卜。你发现了什么？

1. On what number does he jump after 6?
   它跳到6之后会跳到什么数字？

2. On what number does he jump before 10?
   它跳到10之前会跳到什么数字？

3. What number comes after 6?
   6后面是什么数字？

4. What number comes before 9?
   9前面是什么数字？

5. What number is equal to two 3s?
   什么数字等于两个3？

6. If the rabbit stops halfway, where is he?
   如果兔子停在数字线的一半位置，它在哪？
   Write down the number.
   请写下数字。

7. Help the rabbit get back to where he came from. Start at 20.
   Hop one number at a time.
   帮助兔子跳回来。从数字20开始。一次跳一个数字。

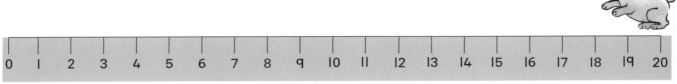

Double up 双倍

Do you like fruit? Which fruits are your favourites?

你喜欢水果吗？哪些水果是你喜欢的？

1. Count the fruit on the first plate. Write down the number.
   数一下第一个盘子里的水果。写下这个数字。
2. Draw **double** this number of fruit on the second plate. Write down the number.
   在第二个盘子中画出双倍的水果。写下新的数字。

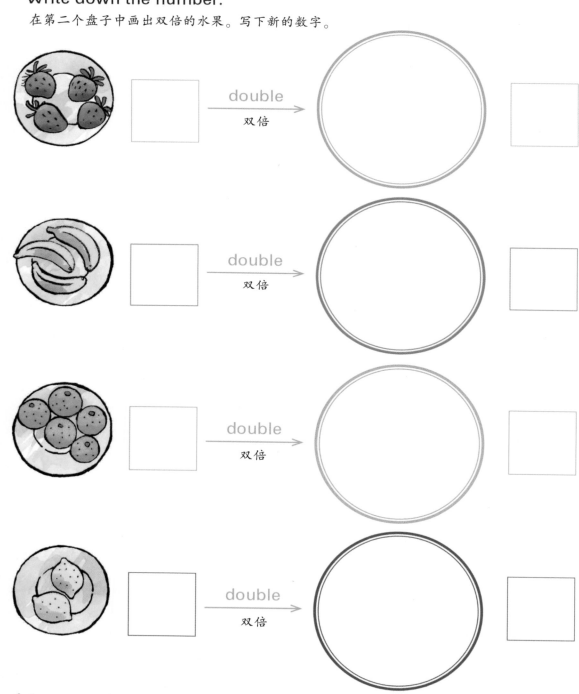

How long does it take?
需要多长时间?

Some things can be done quickly. They take a little time to finish. Some things need to be done slowly. They take more time to finish.

一些事情可以做得很快,只需要一点儿时间来完成。一些事情做起来很慢,需要更多时间来完成。

Put a tick next to each thing that takes a longer time to do.

左右两件事哪种需要更长的时间才能完成,在它们旁边画"√"。

baking a cake
做一个蛋糕

or 或

buttering a slice of bread
在切片面包上涂黄油

walking to school
走路去学校

or 或

driving to school
开车去学校

climbing a tree
爬树

or 或

climbing a mountain
爬山

reading a sign
看一个标志牌

or 或

reading a book
读一本书

washing a cup
洗杯子

or 或

washing a car
洗车

Treasure hunt
寻宝

Collect things from inside your home and from outside.
从家里和外面收集一些物品。

Look for things that are brown.

Look for things that are green.
寻找一些棕色的物品。
寻找一些绿色的物品。

1. Sort the things you found into two groups: a brown group and a green group.
   将你收集的物品分成两组：一组是棕色的物品，一组是绿色的。

2. Draw what you found in the boxes.
   将你找到的物品画在下面方框中。

I found ☐ brown things. Did you find [many] [few] brown things?
我找到___种棕色的物品。你找到了 [很多] [很少] 棕色物品吗？

I found ☐ green things. Did you find [many] [few] green things?
我找到___种绿色的物品。你找到了 [很多] [很少] 绿色物品吗？

Skill: collecting and recording data

30

Serious sewing
复杂的缝纫

Lebo's mom is sewing a new dress for her.
Lebo 的妈妈正在给她做一件新裙子。

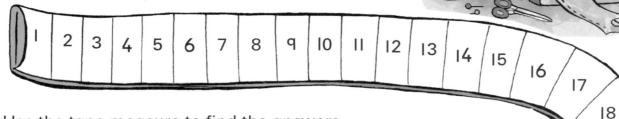

Use the tape measure to find the answers.
Colour in each block as you count.
用卷尺找答案。在你数的数字上涂色。

Start at 12. Count on 4.
从 12 开始，再数 4 个格。

Start at 15. Count on 5.
从 15 开始，再数 5 个格。

Start at 20. Count back 3.
从 20 开始，往回数 3 个格。

Start at 22. Count back 2.
从 22 开始，往回数 2 个格。

Start at 31. Count on 3.
从 31 开始，再数 3 个格。

Start at 11. Count on 6.
从 11 开始，再数 6 个格。

Start at 25. Count on 5.
从 25 开始，再数 5 个格。

Start at 28. Count back 5.
从 28 开始，往回数 5 个格。

Start at 15. Count back 1.
从 15 开始，往回数 1 个格。

Start at 20. Count on 10.
从 20 开始，再数 10 个格。

Start at 20. Count back 10.
从 20 开始，往回数 10 个格。

Munching leaves 咀嚼树叶

Cutie, the caterpillar, is very hungry!
毛毛虫 Cutie 非常饿！

1. Fill in the answers.
   写出答案。

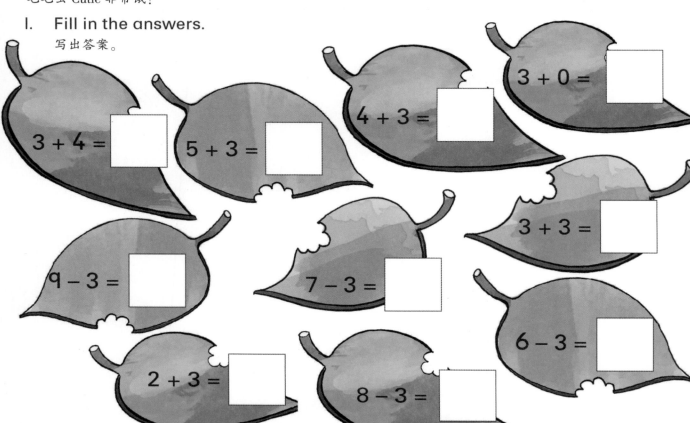

3 + 4 =
5 + 3 =
4 + 3 =
3 + 0 =
9 − 3 =
7 − 3 =
3 + 3 =
2 + 3 =
8 − 3 =
6 − 3 =

2. Cutie ate 4 leaves. He then ate 2 more, then another 2 and then another 2! How many leaves did he eat altogether? Write the number sentence.
   Cutie 先吃了 4 片树叶。他又吃了 2 片，然后又吃了 2 片，最后又吃了 2 片树叶。他一共吃了多少片树叶？请写出算式。

3. There were 8 leaves on the branch. Cutie ate 3. And then another 3. How many leaves are left?
   树枝上有 8 片树叶。Cutie 吃了 3 片，然后又吃了另外 3 片。树枝上还剩多少片树叶？

Spoon it in!
用勺子舀水！

You need: 你需要下列物品：

a yoghurt cup   a spoon   a bowl of water   a tin
一个酸奶杯   一个勺子   一碗水   一个罐子

1. Take a guess! Write your answer.  估计数
   猜测一下，并写出你的答案。

   How many spoons of water will fill your yoghurt cup?
   盛满酸奶杯要舀入几勺水？

   How many tins of water will fill your yoghurt cup?
   盛满酸奶杯要倒入几罐水？

2. Work it out! Write your answer.  实际测量
   实际操作一下，并写出你的答案。

   Spoon the water into your yoghurt cup. Count how many spoons of water
   用勺子将水舀进酸奶杯。数一下几勺水能把杯子
   盛满。
   it takes to fill the cup. The cup holds spoons of water.
   杯子里盛了___勺水。

3. Pour the water back into the bowl.
   把水倒进碗里。

   Use the tin to scoop the water into your yoghurt cup. Count how many tins
   用罐子将水舀入酸奶杯。数一下几罐水能把杯子倒满。
   of water it takes to fill the cup. The cup holds tins of water.
   杯子里盛了___罐水。

Sports day 运动会

The children are doing the sack race at sports day. What fast jumpers!
孩子们在运动会上进行套袋跳赛跑。他们跳得多快啊!

1. Who is coming 1st? _____
   谁是第1名?

2. Who is coming 3rd? _____
   谁是第3名?

3. Max is in _____ position.
   Max 是第___名。

4. Judy is in _____ position.
   Judy 是第___名。

5. Brian is in _____ position.
   Brian 是第___名。

6. Jaco is in _____ position.
   Jaco 是第___名。

7. Is Edward in 4th or 5th position? _____
   Edward 是第4名还是第5名?

8. Is Lebo in 7th or 8th position? _____
   Lebo 是第7名还是第8名?

START 起点

FINISH 终点

Skill: ordering numbers

34

Lots of spots
很多的斑点

What a lot of spots these ladybirds have!
这些瓢虫身上有好多斑点!

1. Count the spots on the ladybirds and write number sentences. The first one has been done for you.
   数一下瓢虫身上的斑点,并将算式写下来。请参考第一行的例子。

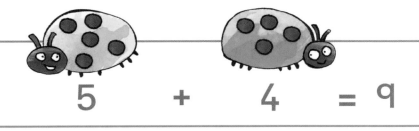

5 + 4 = 9   nine

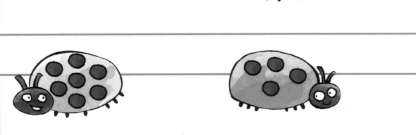

2. Write the names of the numbers. Do you know them all? Ask for help if you need it.
   写出数字的英文单词。你会写吗? 如果有不会写的字可以请求帮助。

Money fun 存钱的乐趣

Mandla has been saving his pocket money to buy a soccer ball.
Mandla 一直在存零用钱,想去买一个足球。

Here are the coins in Mandla's piggy bank:
下面的硬币是 Mandla 储蓄罐中的硬币：

1. Help Mandla to add up how much money he already has.
   帮 Mandla 算一下他一共存了多少钱。

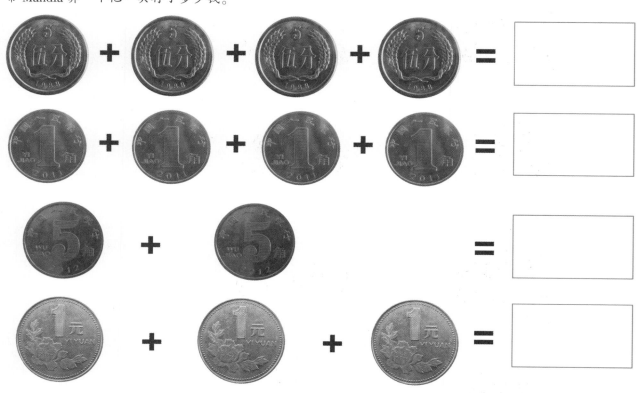

2. Last week Mandla spent 3 yuan on sweets before he saved his pocket money. His pocket money was 10 yuan. How much did he save?

   上周 Mandla 在将钱存起来之前,他花了 3 元钱买糖。
   他有 10 元零用钱。他存起来多少钱?

Emma's new crayons
Emma 的新蜡笔

Emma got four new boxes of crayons for her birthday. What a lot of crayons!

Emma 得到 4 盒蜡笔作为生日礼物。好多蜡笔呀!

1. Help Emma **guess** how many crayons are in each box. Then **count** to see how many crayons there really are.

   帮 Emma 猜一下每个盒子中有多少支蜡笔。然后数一下每个盒子中实际的蜡笔数。

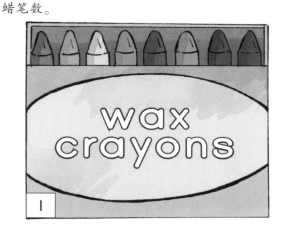 

estimate 估计数   count 实际数   estimate 估计数   count 实际数

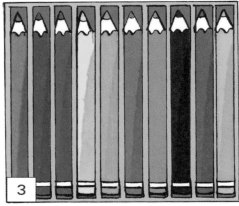 

estimate 估计数   count 实际数   estimate 估计数   count 实际数

2. Which box has the fewest crayons?
   哪个盒子中蜡笔数量最少?

3. Which boxes have an equal number of crayons? and
   哪两个盒子蜡笔数量相同?

Royal mats
城堡毯子

Princess Lebo lives in a castle with long passages. She wants some new mats to brighten up the castle.

Lebo 公主住在一座城堡中,城堡中有长长的走廊。她想用一些新的毯子来装饰城堡。

1. Look at the pattern on the castle mat. Which shape comes first in the pattern? Draw it.

   看下面毯子的花纹。第一个花纹是什么形状? 将它画下来。

   Which shape comes second in the pattern? Draw it.
   第 2 个花纹是什么形状? 将它画下来。

   Which shape comes third in the pattern? Draw it.
   第 3 个花纹是什么形状? 将它画下来。

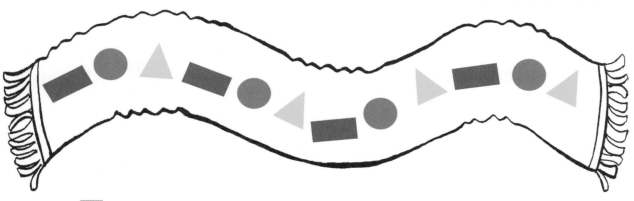

2. Use  and to draw a pattern on this mat.

   用 和 在这条毯子上画个有规律的花纹。

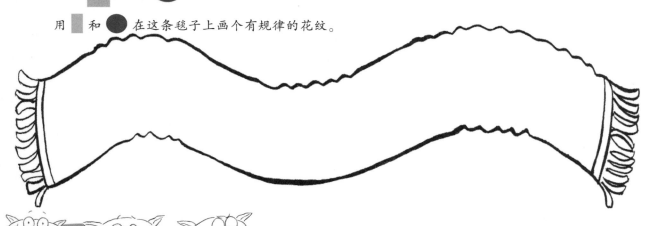

Clever guesses
聪明的猜测

Here's some guessing fun for you. 下面是一些猜猜乐游戏。
Remember to estimate first and then count. 记住先估计一下,再具体数一下。

How many are there in the picture? 在图片中,下面列出的各有多少?

	estimate 估计数	count 实际数
cars 汽车		
windows 窗户		
trees 树		
people 人		
buildings 建筑物		

Number gym

数字游戏

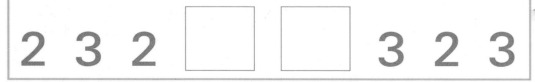

Let your brain do some exercise!

让你的头脑做一些游戏！

1. Can you see the patterns? Fill in the missing numbers.

   看看下面的数列。将缺少的数字填上。

   | 2 | 3 | 2 | ☐ | ☐ | 3 | 2 | 3 |

   | 7 | 8 | 9 | ☐ | 8 | 9 | 7 | ☐ | 9 |

   | 5 | 1 | 1 | 5 | 1 | 1 | ☐ | 1 | 1 | 5 | ☐ | 1 |

2. Fill in the missing numbers.

   将缺少的数字填上。

40

Shapes and things
形状和物品

Can you see shapes in the objects around you?
你能看出这些身边经常看到的物品是什么形状吗?

1. Draw different colour lines to match the shapes to the objects.
   用不同颜色的线将相同形状和物品相连。
2. Say the names of the shapes.
   说出这些形状的名称。

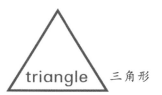
triangle 三角形

sphere 球形

rectangle 长方形

prism 棱柱体

sphere 球形

circle 圆形

All aboard
请上车

The train can only leave the station when it has been shown a green flag.

只有出示绿色旗子之后，火车才能开动出站。

Write the answers in the flags. Colour the flags green.

将算式答案写在旗子上。将旗子涂成绿色。

4 + 2 + 2 =

5 + 2 + 4 =

8 + 2 + 2 =

9 + 3 + 3 =

Buckets and bowls 桶和碗

You need: 你需要下列物品：

water
水

a large bowl
1个大碗

a jug
1个壶

a bucket
1个桶

1. Take a guess! Write your answer.  估计数
   猜测一下，写出你的答案。

   How many bowls will fill the bucket of water?
   几碗水可以装满 1 桶？

   How many jugs of water will fill the bucket?
   几壶水可以装满 1 桶？

2. Work it out! Write your answer.  实际数
   实际操作一下，写出你的答案。

   Pour bowls of water into the bucket until it is full. Count as you go.
   用碗向桶里倒水，直到将桶装满。一边倒水一边计数。
   How many bowls of water did you use to fill the bucket?
   要用多少碗水能装满 1 桶？

   The bucket holds ☐ bowls of water.
   1 桶能装____碗水。

   Pour jugs of water into the bucket until it is full. Count as you go.
   用壶向桶里倒水，直到将桶装满。一边倒水一边计数。
   How many jugs of water did you use to fill the bucket?
   要用多少壶水能装满 1 桶？

   The bucket holds ☐ jugs of water.
   1 桶能装____壶水。

Same, same 相同,相同

Many things are the same on both sides. They are symmetrical.
很多事物左右两边是相同的。也就是说,它们是对称的。

You need:
你需要下列物品:

string   a wooden spoon   a book   an apple   a T-shirt
绳子   1个木头勺子   1本书   1个苹果   1件T恤

Lay the string down the middle of the spoon, book and T-shirt.
将绳子放在勺子、书和T恤的中间部分。

1. Is each side the same? Are the objects symmetrical?

   勺子左右两边相同吗?它是对称的吗?

2. Ask an adult to help you cut the apple in half. Is each side the same? Is the apple symmetrical?

   让家中成年人帮忙将苹果切成两半。它的两半相同吗?这个苹果是对称的吗?

3. Can you guess what these objects are? 你能猜出这些是什么物品吗?
   Draw the other half of each one. 将它们的另一半画出来。

4. Find something at home that is **not** symmetrical.

   找出家中一些不对称的物品。

Show of hands 伸出双手

These children are copying their teacher.
这些孩子在跟着老师做。

1. Count the hands in the picture. ☐

   数一下图片中共有几只手。

   Count the fingers in the picture. Count in 5s. ☐

   数一下图片中共有几根手指。5个5个地数。

2. Count the hands in the picture. ☐

   数一下图片中共有几只手。

   Count the fingers in the picture. Count in 10s. ☐

   数一下图片中共有几根手指。10个10个地数。

3. Count the hands in the picture. ☐

   数一下图片中共有几只手。

   Count the fingers in the picture. Count in 2s. ☐

   数一下图片中共有几根手指。2个2个地数。

I spy 我来观察

Some edges are curved.
一些物品的边是弧形的。

Some edges are straight. _____
一些物品的边是直线的。

Some things roll.
一些物品能滚动。

Some things slide.
一些物品能滑动。

Complete the sentences. Use the green and blue words above and the red words in the boxes below to help you.
完成下面的句子。可以借助上面绿色的、蓝色的单词和下面方框中红色的单词。

circle	prism	triangle	sphere
圆形	棱柱	三角形	球形

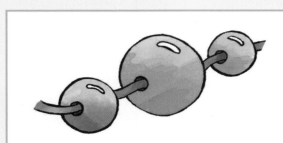

I have _____ edges.
　　　　　边,边缘
I _____ .
I am a _____ .

I have _____ edges.
　　　　　边,边缘
I _____ .
I am a _____ .

I have _____ edges.
　　　　　边,边缘
I _____ .
I am a _____ .

I have _____ edges.
　　　　　边,边缘
I _____ .
I am a _____ .

How many marbles?

有多少颗弹珠?

Jaco and his friends like to play marbles. Sometimes they win marbles. Other times they lose marbles. Help them work out how many marbles they have now.

Jaco 和他的朋友喜欢玩弹珠。有时他们能赢一些弹珠。有时会输一些弹珠。帮他们算一下现在他们有多少颗弹珠。

1. **Estimate** first, then **count**. Fill in your answers.
   先估计一下,然后再实际计数。写出你的答案。

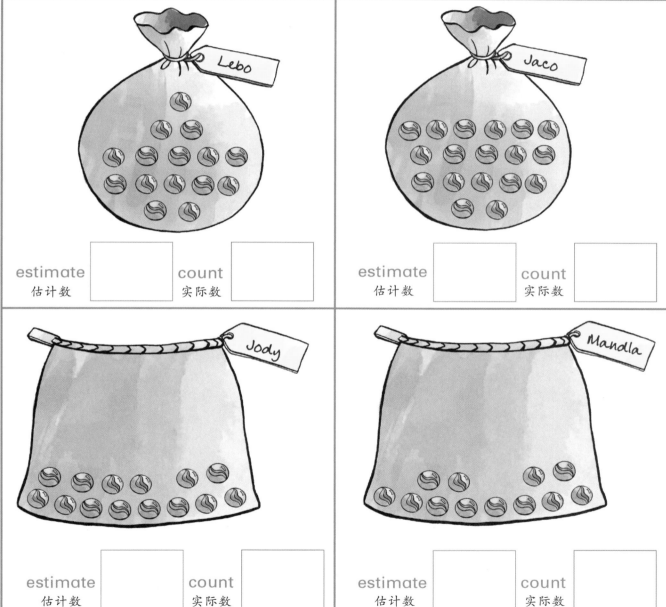

2. Who has the most marbles? _____
   谁的弹珠最多?

3. Who has the fewest marbles? _____
   谁的弹珠最少?

Doubling with dominoes
骨牌加倍

1. Use your stickers to double the number of dots on each domino.
   用书后的贴纸将每个骨牌的点数加一倍。
2. How many dots are on each domino now?
   现在每个骨牌有多少点数。

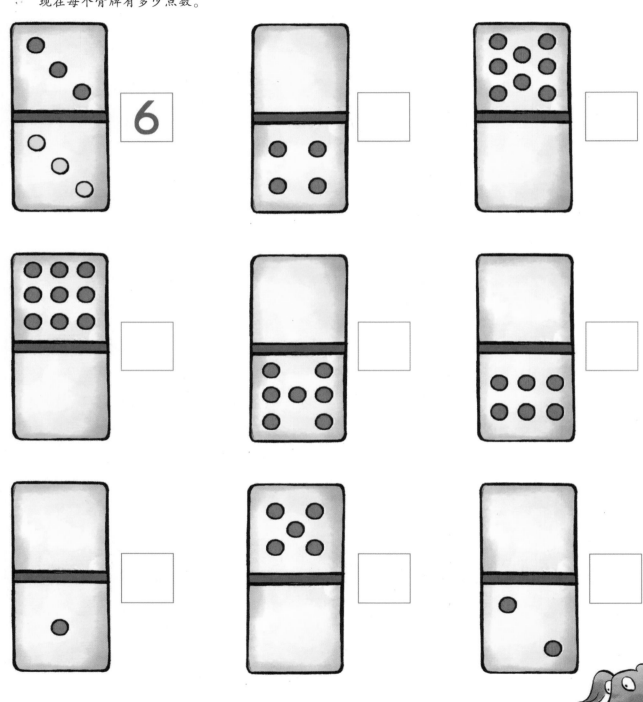

Lebo's beady eye
Lebo 雪亮的眼睛

Lebo has guessed how many beads there are in each jar.
Lebo 猜测了每个罐子中有多少颗小珠子。

1. Without counting the beads, match Lebo's guesses to the jars.
   在不数小珠子的情况下,将 Lebo 猜的数与相应的罐子相连。

2. Now count the beads and write the number in the boxes.
   现在数一下罐子中小珠子的数量,并将答案写在旁边的方框中。

Lebo's guesses
Lebo 的猜测数

3. Which jar is closest to Lebo's estimate? jar
   哪个罐子中小珠子的实际数量与 Lebo 的猜测最接近?

At school
在学校

Here is another number line:
下面是一条数字线：

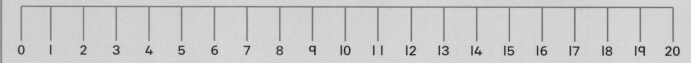

1. Use the number line to work out the answers. First look at the examples.
   用数字线算出答案。先看一下例子。

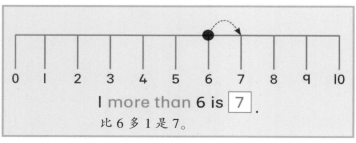

1 more than 6 is 7.
比 6 多 1 是 7。

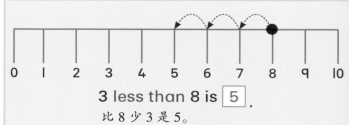

3 less than 8 is 5.
比 8 少 3 是 5。

1 more than 12 is ☐.
比 12 多 1 是 ___。

3 more than 9 is ☐.
比 9 多 3 是 ___。

3 more than 5 is ☐.
比 5 多 3 是 ___。

5 less than 20 is ☐.
比 20 少 5 是 ___。

1 less than 19 is ☐.
比 19 少 1 是 ___。

4 less than 10 is ☐.
比 10 少 4 是 ___。

1 more than 11 is ☐.
比 11 多 1 是 ___。

4 more than 5 is ☐.
比 5 多 4 是 ___。

Look at Stinky and the children.
看小狗 Stinky 和这些孩子。

2. How many eyes do they have altogether? Count in 2s. ☐
   他们一共有多少只眼睛？2 个 2 个地数。

Toy box tally

给玩具箱做标记

Emma's mother says she must give away some of her toys. Emma can't decide which things to give away. Can you help her find out how many of each type of toy she has?

Emma 的妈妈说要把她的一些玩具扔掉。Emma 不能决定扔掉哪些玩具。你能帮她分类看看每种玩具有多少个吗?

1. Put stickers in the table to show how many of each type of toy Emma has.

   将 Emma 每种玩具的个数用书后贴纸标记在表格中。

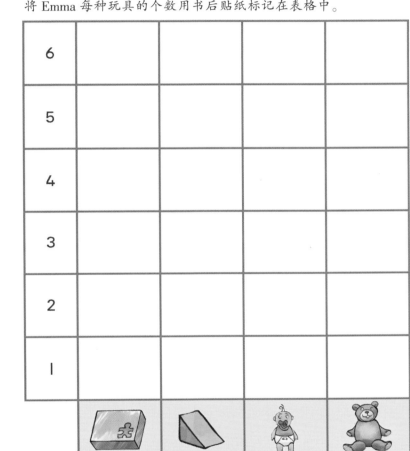

2. How many does Emma have of each type of toy?

   Emma 每种玩具有多少?

   puzzles 拼图 ☐   blocks 积木 ☐   dolls 玩偶 ☐   soft toys 毛绒玩具 ☐

3. Of which toy does Emma have the most? _____

   Emma 哪种玩具数量最多?

4. Of which toy does she have the fewest? _____

   她哪种玩具数量最少?

Let's go shopping

让我们去购物

The sweet shop is having a sale. Jody visits the shop with her pocket money. Help her decide what to buy.

糖果店正在营业。Jody 带着零用钱到了这家商店。帮助她挑选糖果。

1. Find the correct sticker.
   找到正确的贴纸。

   a. Which item costs the least amount of money? _____
      哪一种糖果花费钱数最少?

   b. Which item costs the most? _____
      哪一种糖果花费钱数最多?

   c. Which item costs 5jiao? _____
      哪一种糖果花费 5 角?

   d. Jody has 8jiao in her purse. Which items can't she buy? _____
      Jody 钱包里有 8 角。哪些糖果她买不了?

   e. Jody buys something. She gets 9fen change.

      What did she buy? _____
      Jody 买了一些糖果。找了 9 分零钱。她买的是什么?

2. If you had 3yuan in your purse, what could you afford to buy from the sweet shop? Use your stickers or draw the items.
   如果你钱包里有 3 元，你能从糖果店买什么？可以使用贴纸或者画出这些糖果。

3. How many marshmallows can Jody and Emma buy with 2yuan? How many will they each get?
   Jody 和 Emma 用 2 元可以买多少棉花糖? 他们每人可以分多少?

Winter woollies 冬天的毛线物品

It's washing day. Dad is hanging winter clothes on the line.
在洗衣日,爸爸将冬天的衣物晾在绳上。

1. Count the washing on the line. How many items of clothing are there?

   Count from the first sock. ☐

   数一下绳上晾着的衣物。每种衣物有多少件?从第一行的短袜开始算起。

2. Check yourself. Count backwards from the last sock. 自己检查一下。从最后一只袜子倒数回来。答案一样吗?

   Is your answer still the same? _____

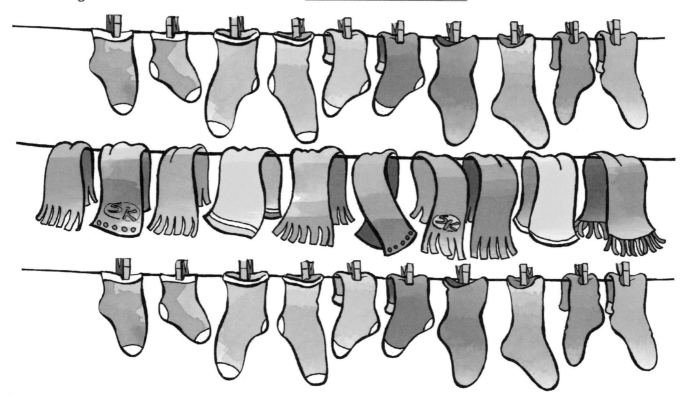

3. Count in 2s and complete the number pattern. 2个2个地数,完成这个数列。

4. Count in 3s and complete the number pattern. 3个3个地数,完成这个数列。

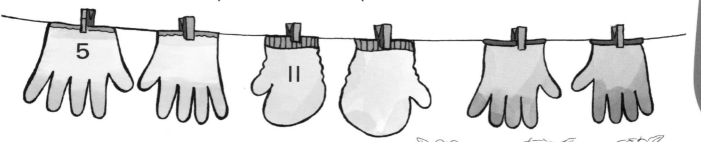

Missing numbers
缺少的数字

1. Fill in the missing numbers.
   将缺少的数字补齐。

1	2		4		6	7	8	9	10
11		13		15	16		18	19	
21	22	23	24	25	26	27	28	29	30
31	32	33	34	35	36	37	38	39	40
41	42	43	44	45	46	47	48	49	50
51	52	53	54	55	56	57	58	59	60
61	62	63	64	65	66	67	68	69	70
71	72	73	74	75	76	77	78	79	80
81	82	83	84	85	86	87	88	89	90
91	92	93	94	95	96	97	98	99	100

2. What patterns can you see on the number block?
   在数字表格中你能发现什么规律?

3. What is 1 more than 55? Colour the block red.
   比 55 多 1 的数字是多少? 将那个方格涂成红色。

   What is 5 more than 54? Colour the block blue.
   比 54 多 5 的数字是多少? 将那个方格涂成蓝色。

   What is 2 more than 21? Colour the block green.
   比 21 多 2 的数字是多少? 将那个方格涂成绿色。

   What is 4 less than 66? Colour the block yellow.
   比 66 少 4 的数字是多少? 将那个方格涂成黄色。

   What is 3 more than 35? Colour the block orange.
   比 35 多 3 的数字是多少? 将那个方格涂成橙色。

4. What is the biggest number in the block? Circle it.
   方格中最大的数字是多少? 将它圈出来。

Tropical birds 热带鸟儿

These pretty birds are sitting on the branches of a tree.
这些漂亮的鸟儿栖息在一棵树的树枝上。

Find their friends so that they are sitting in order from 1st to 20th. Use your stickers to place the birds on the branches.
它们的朋友栖息在 1~20 之间，找到它们。用贴纸标记出鸟儿在树枝上的位置。

Computer fun
计算机游戏

Follow the instructions on the computer screens to help Mandla find the answers.

按照计算机屏幕上的指令帮助 Mandla 找到答案。

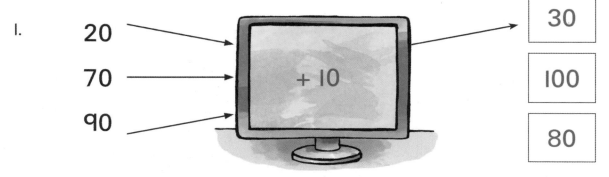

1. 20 → +10 → 30
   70 → +10 → 100
   90 → +10 → 80

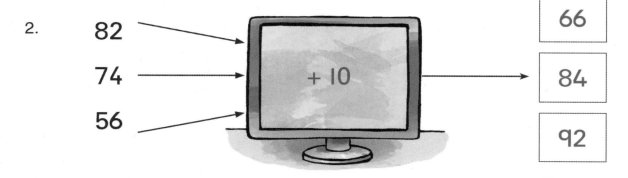

2. 82 → +10 → 66
   74 → +10 → 84
   56 → +10 → 92

3. 10 → −10 → 90
   30 → −10 → 0
   100 → −10 → 20

4. 74 → −10 → 13
   49 → −10 → 39
   23 → −10 → 64

Hands on
用手测量

> A grown-up needs to help you with this activity.
> 可以找个成年人帮你完成这个练习。

Mandla uses his hands to measure. You can use your hands like this:
Mandla 用手进行测量。你也可以像这样用手测量:

1. Stretch your fingers wide.
   伸展开你的手指。

2. Estimate how many hands it will take to measure the height of each thing on this page. Write down the number of hands.
   估计一下家中下列物品用手测量是几个手高度。写下答案。

3. Now go and measure the height of these items in your own home using your hands. Write down the number of hands.
   现在用手实际测量一下这些物品的高度。写出是几个手高度。

estimate 估计数

measure 实际测量数

estimate 估计数

measure 实际测量数

estimate 估计数

measure 实际测量数

4. Which is the tallest? _____
   哪个物品最高?

5. Which is the shortest? _____
   哪个物品最矮?

6. Draw a star next to your closest estimate.
   在与你估计数最接近的物品旁画一个星星。

Skill: measuring

57

Peek-a-boo! 躲猫猫!

Which number is hiding in the number square?
哪个数字隐藏在下方的数字方块中?

1. Work out the answers. 算出答案。

Half of 40	☐	Half of 30	☐	Double 9	☐
40 的一半		30 的一半		9 的 2 倍	
Half of 10	☐	Double 20	☐	Double 8	☐
10 的一半		20 的 2 倍		8 的 2 倍	
Half of 12	☐	Double 5	☐	Double 7	☐
12 的一半		5 的 2 倍		7 的 2 倍	
Half of 6	☐	Double 4	☐		
6 的一半		4 的 2 倍			

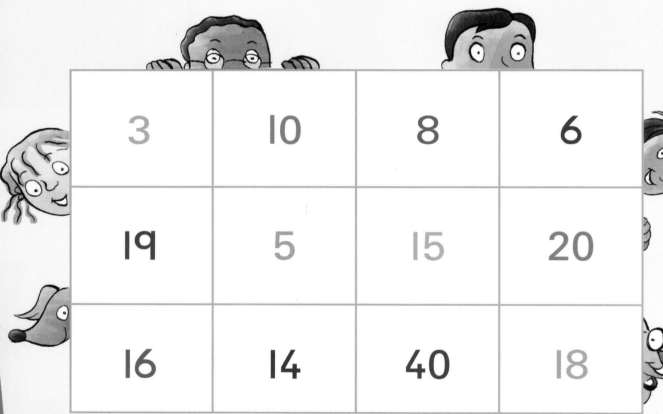

3	10	8	6
19	5	15	20
16	14	40	18

2. Colour the squares in the number block to match the numbers with your answers. Which number is hiding?
   将数字方块中与你的答案相匹配的数字方框涂上颜色。哪个数字隐藏在其中?

3. Draw a circle around the biggest number in the number block.
   在数字方块中最大的数字上画圈。

4. Draw a line under the smallest number in the number block.
   在数字方块中最小的数字下画横线。

Body snap! 身体快照!

Can you make both sides the same? 你能让身体的两边相同吗?

1. Draw the other half of Mandla. 画出 Mandla 另一半身体。

Draw the same body parts on the right. Make sure that you match the patterns on the clothes too.

画出相对应的右边身体。确保衣服上的图案也对应。

2. What number does Mandla have on his shirt? ☐

Mandla 衣服上的数字是多少?

Jody the witch
女巫 Jody

Help Jody to cast her magic spell. The numbers on each cauldron must add up to the number on the hat for Jody's spell to work. Colour each hat as you count.

帮助 Jody 施展魔法。需在每个大锅上加 1 个数字使其总和与帽子上的数字相同,才能让 Jody 的魔法咒语生效。一边计算一边给每顶帽子涂色。

1. Write the correct number on the cauldron.
   在锅上填写正确的数字。

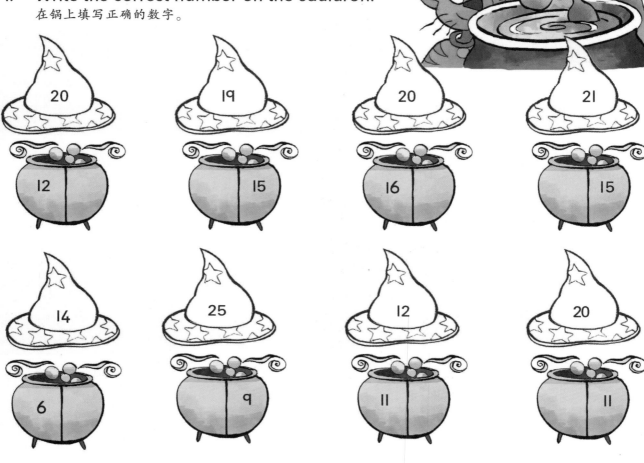

2. Complete the number sentences. Use colour to match the hats and the number sentences. 完成下列算式。用不同的颜色使帽子与算式相匹配。

12 + 8 = ☐   15 + ☐ = 21   12 − ☐ = 11

☐ − 4 = 15   6 + ☐ = 14   20 − 11 = ☐

20 − 16 = ☐   25 − 9 = ☐

Weigh to go!
称重！

Mandla is using a balancing scale and marbles to weigh some of his things.
Mandla 正在用天平和弹珠为他的一些物品称重。

1. Write down how many marbles each object weighs.
   写出每种物品重量等于几个弹珠。

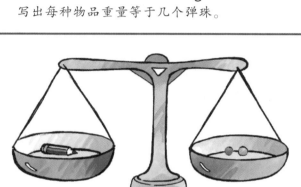

The pencil weighs ⬜ marbles.
铅笔的重量为___个弹珠。

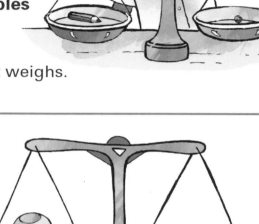

The ball weighs ⬜ marbles.
球的重量为___个弹珠。

The toothbrush weighs ⬜ marbles.
牙刷的重量为___个弹珠。

The shoe weighs ⬜ marbles.
鞋子的重量为___个弹珠。

2. Which object is the heaviest? _____
   哪个物品最重？

3. Which object is the lightest? _____
   哪个物品最轻？

4. How do you know? _____
   你怎么算出最重和最轻的物品的？
   _____

Where is it?
在哪里？

What is behind you?
What is in front of you?
哪种物品在你后面？
哪种物品在你前面？

You will need: 你需要下列物品：

a coin   an eraser   a pencil   a small box   a crayon
1个硬币   1块橡皮   1支铅笔   1个小盒子   1支蜡笔

1. Put the small box on the x.
   将小盒子画在 X 的位置上。

X

2. Put the behind the box. 将橡皮放在盒子后面。

3. Put the in front of the box. 将硬币放在盒子前面。

4. Put the behind the box. 将蜡笔放在盒子后面。

5. Put the in front of the box. 将铅笔放在盒子前面。

6. What shape is the box? _____
   盒子是什么形状的？

Sorted! 分类！

Our food comes in different containers.
我们的食物装在不同的容器中。

1. Collect 15 items of food from your kitchen. The food must be in:
   从你的厨房中找出 15 种食物。要求这些食物必须是装在下列容器中。

    boxes 盒子   cans 易拉罐   jars 罐子

2. Draw pictures to show how many you collected. Start drawing from the bottom of the table.
   在相应的方框中画出你收集了多少食物。从表格的底部开始画。

10			
9			
8			
7			
6			
5			
4			
3			
2			
1			
	 boxes 盒子	cans 易拉罐	jars 罐子

3. How many do you have? 下列容器装的食物你各有几种。

    boxes 盒子 ☐   cans 易拉罐 ☐   jars 罐子 ☐

4. What have you collected the most of? _____
   哪种容器的食物你收集的最多？

Poles and ladders
杆子和梯子

Have fun playing this firefighter game on your own or with someone else.
自己玩儿或者和别人一起玩儿消防员的游戏。

You will need a die and a counter for each player. Use the stickers from the sticker page to make counters.

每个游戏者需要一个骰子和一个计数器。从贴纸页上取一个贴纸来计数。

Climb up the ladders.
Slide down the poles.

顺着梯子爬上去。
顺着杆子滑下来。

Rules for playing on your own
- Start by rolling the die.
- Move your counter the same number of blocks as on the die.
- You are the winner when you reach number 100.

自己玩儿的游戏规则：
- 开始掷骰子。
- 按照骰子数向前走几格。
- 到达 100 时就是胜利。

Rules for playing with other people
- Roll the die.
- The person who throws the highest number on the die, starts the game.
- Take turns to roll the die and move your counters.
- The first person to reach 100 is the winner.

和别人一起玩儿的游戏规则：
- 掷骰子。
- 谁掷的点数最大，谁先开始游戏。
- 轮流掷骰子，然后向前移动所掷的数。
- 第一个到达 100 的人获胜。

Fancy flowers
美丽的鲜花

Jaco loves gardening. Help him to plant the beautiful flowers in rows. Use the stickers.

Jaco 喜欢园艺。帮助他按排种植美丽的花儿。用书中的贴纸完成。

1. Jaco has 12 white daisy plants. Plant them in 3 equal rows.
   Jaco 有 12 棵白色的雏菊幼苗。 将它们平均种在 3 排中。

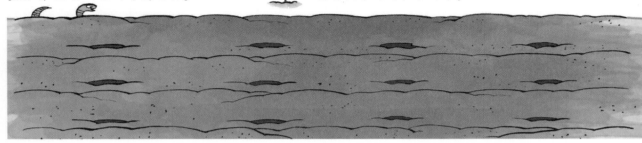

How many plants are there in each row? ☐
每一排能种多少棵雏菊?
How many plants are left over? ☐
剩下多少棵雏菊?

2. He has 16 red rose plants. Plant them in 3 equal rows.
   他有 16 棵红玫瑰幼苗。 将它们平均种在 3 排中。

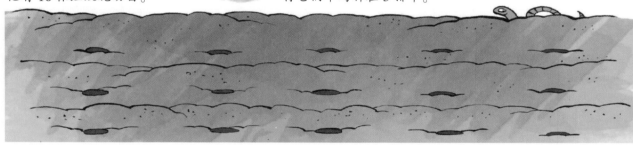

How many plants are there in each row? ☐
每一排能种多少棵玫瑰?
How many plants are left over? ☐
剩下多少棵玫瑰?

3. He has 11 pink tulip plants. Plant them in 2 equal rows.
   他有 11 棵粉色郁金香幼苗。 将它们平均种在 2 排中。

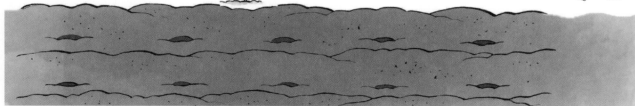

How many plants are there in each row? ☐
每一排能种多少棵郁金香?
How many plants are left over? ☐
剩下多少棵郁金香?

Keep counting
继续数数

Can you help Mandla work out these sums?
你能帮助 Mandla 算出下列物品总数吗?

   =

toes
脚趾

 −  −  −  =

socks
袜子

 =

hands
手

 −   − =

gloves
手套

 +  +  +  +  =

balloons
气球

67

High five 伸手击掌

Let's count fingers!
让我们数数手指！

1. How many fingers do you have on one hand? ☐
   一只手有几根手指？

2. How many fingers do you have on both hands? ☐
   两只手有几根手指？

3. How many fingers are in the picture altogether? Count in 10s.

   Complete the number sentence. 图中一共有多少根手指？10个10个地数。完成下列算式。

   ☐ + ☐ + ☐ + ☐ + ☐ = ☐

4. Write the answers. Use the number line to help you. 用数字线帮你写出算式答案。

   20 − 5 = ☐   11 + 5 = ☐

   5 + 5 = ☐   9 − 5 = ☐

   11 − 5 = ☐   15 + 5 = ☐

   6 − 5 = ☐   8 − 5 = ☐

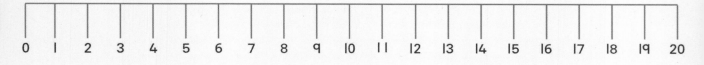

What is the pattern?

这是什么规律？

Help Emma and Lebo to make their pretty necklaces by filling in the missing numbers.

帮助 Emma 和 Lebo 补齐缺少的数字，做成漂亮的项链。

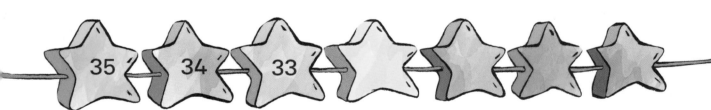

Happy birthday!

生日快乐！

Do you know when your next birthday is?

你知道自己下一次的生日是什么时间吗?

1. Find your birthday month on a calendar.
   在日历上找出你生日所在的月份。
2. Copy the name of the month onto the calendar below.
   将月份的名称写到下面的日历中。
3. Write the year.
   写出年份。
4. Copy the dates.
   写出日期。
5. Draw a party hat on the date of your birthday.
   在你生日当天的空格中画一顶生日帽。

Month: 月份　　　　　　　　　　　　　　　Year: 年份

Sunday 星期日	Monday 星期一	Tuesday 星期二	Wednesday 星期三	Thursday 星期四	Friday 星期五	Saturday 星期六

6. When is your birthday? _____
   你的生日是什么时间?

7. On what day does your birthday fall this year? _____
   今年你的生日是星期几?

Happy birthday, Gran!

奶奶,生日快乐!

**Jaco's gran is turning 70.
Do they have enough candles?**

Jaco 的奶奶快 70 岁了。他们准备的蜡烛够吗?

1. **Estimate** how many candles there are.

   估计一下一共有多少根蜡烛。

2. Now **count** the candles. Draw circles around the candles to make groups of 10. How many are there?

   现在数一下蜡烛。每 10 根画一个圈。一共有多少根?

3. How many more candles do they need for the cake?

   蛋糕上还需要多少根蜡烛?

4. Count in **5s** on the number line. Start at 0.

   从 0 开始,用数字线 5 个 5 个地数。

   How many counts to get to 50?

   多少次能数到 50?

   How many counts to get to 70?

   多少次能数到 70?

   How many counts to get to 100?

   多少次能数到 100?

   0 5 10 15 20 25 30 35 40 45 50 55 60 65 70 75 80 85 90 95 100

Hide and seek
捉迷藏

Mandla is trying to find his friends. They are hiding. Can you help him?

Mandla 正在找他的朋友们。他们藏起来了。你能帮助他吗?

1. Find Mandla's friends.
   找 Mandla 的朋友们。

2. Find: 寻找下列形状：
    - a rectangle 长方形
    - a sphere 球形
    - a triangle 三角形
    - a circle 圆形
    - an oval 椭圆形
    - a square 正方形
    - a prism 棱柱体

3. Choose an ending for each of the sentences.
   用下列方框中的文字来补齐下列句子。

   | in the bush 在草丛中 | behind the tree 在树后面 | behind the bush 在草丛后面 |
   | next to Mandla 在 Mandla 旁边 | under the bench 在凳子下面 | on the wall 在墙上 |

   Jody is _____.

   Lebo is _____.

   Emma is _____.

   Ravi is _____.

   Jaco is _____.

   Honey is _____.

Shop till you drop 逛逛商店

What can you afford? Read the price tags and count the money.
你可以购买下列物品吗？看一下标签并数一下有多少钱？

1. Match the money with a toy. 将玩具和相同的钱数连线。

   a. 1元 + 1元

   b. 1元 + 1元 + 5角 + 5角

   c. 1元 + 1元 + 1元 + 1元

   d. 5元 + 5元

   e. 10元 + 10元

   car – 4yuan
   ball – 3yuan
   rabbit – 10yuan
   lollipop – 2yuan
   bear – 20yuan

2. If you had 20yuan to spend on each item, how much change would you get each time? 如果你用20元分别买下列物品，每种找回多少钱？

   rabbit _____   ball _____   bear _____
   兔子   球   熊

   car _____   lollipop _____
   小汽车   棒棒糖

3. Complete the sentences. Use words not numbers. 完成下列句子，用单词而不是用数字。

   a. The 🚗 plus the 🍭 cost _____ yuan altogether.
      小汽车和棒棒糖一共花___元。

   b. The ⚽ plus the 🚗 cost _____ yuan altogether.
      球和小汽车一共花___元。

   c. The ⚽ plus the 🚗 plus the 🍭 cost _____ yuan altogether.
      球、小汽车和棒棒糖一共花___元。

   d. The 🐻 plus the 🐰 cost _____ yuan altogether.
      小熊和小兔子一共花___元。

Does it belong?

物品归类？

1. Colour the pictures of things that roll.
   将图片中能滚动的物品涂色。

2. Colour the pictures of things that slide.
   将图片中能滑动的物品涂色。

3. Colour the shapes with curved edges.
   将图片中弧形边的图形涂色。

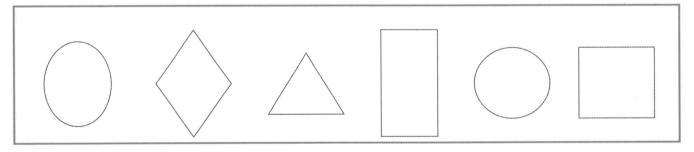

4. Colour the shapes with straight edges.
   将图片中直线边的图形涂色。

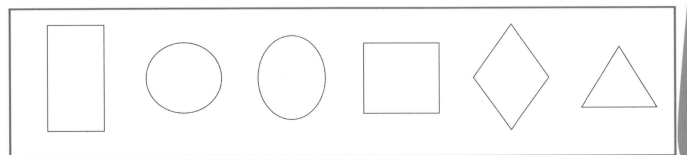

Notes
注释

下面列出帮助孩子训练的一般准则：
- 用真实的事物帮助孩子理解概念。如，2个叉子加上2个叉子等于4个叉子。
- 在开始练习前，你需要准备游戏所需的材料。
- 一直和孩子一起阅读提示。
- 年幼的孩子们注意力集中的时间很短，所以让孩子一次只做一两页即可。

P2
先让孩子试着找到事物的规律。如，叉子、勺子、叉子、勺子。这种练习可以帮助孩子识别和运用多达4种形状和4种颜色的规律。帮助孩子掌握书中的数字规律，包括从3倒数到1。

P3
在这个游戏中，孩子可以认识2种三维物体：棱柱体(盒子)和球体(球)。首先，在家中寻找像球和盒子这些形状的物品。在孩子完成这项练习的时候，跟他谈一谈图片中哪些物品的边是弧形的（如球或者球形体），哪些物品的边是直线的(如棱柱体或盒子)。

P4
让孩子从左向右数，并理解数字从左到右是增加的。青蛙的每一次跳跃数字都增加1。向孩子介绍如何在数字线的计数上做减法。
答案：(1) 2, 3, 1 (2) 0, 1 (3) 4, 5 (4) 2, 3

P5
先让孩子将身边的一些物品进行分类。然后看一下书中图片，如果他能够将蜡笔用颜色来进行分类，那么这种分类是最理想的。让他用贴纸在方框中将蜡笔进行分类，再将每个方框中蜡笔的数量数一下，会发现绿色的蜡笔数量是最多的。这样也可以练习数数，问问他蜡笔一共有多少支。

P6
这个游戏的重点在于区分24小时中不同的时间段：早上、中午和夜晚。首先和孩子一起读一下句子要求。将句子补充完整后，和孩子再一起读一下，并且和孩子一起讨论一下这些事情是不是他平时也会做的事情。
答案：morning；night；afternoon

P7
这个游戏可以帮助孩子练习数数，并且树立"更多"的概念。必要的时候，提醒一下孩子别忘了图片中主人公手中抱的泰迪熊。
答案：(1)5, 8 (2)10 (3)Lebo

P8
这个游戏中，孩子学习的是重量的概念。他需要考虑到不同的事物或人的重量，再与自身来比较，然后选出这些事物中最重和最轻的。
答案：(2)更重：成年人、扶手椅、小汽车、炉子、大象
更轻：砖块、铅笔、一花瓶的花、垫子
(3,4)小汽车是所有物品中最重的，铅笔是最轻的。

P9
孩子需要发现每行图形或数字中存在的规律。然后再将每行补充完整。
答案：(1)蓝圆形 红三角，蓝圆形 红三角 (2)绿星星 紫星星 红菱形 紫菱形 (3)橙圆形 红心形，黄方形 橙圆形 红心形 (4)2, 3, 4, 5；(5)4, 3, 2

P10
这个游戏帮助孩子阅读和使用日历。孩子可能无法读出日历上所有的数字，但他应该会使用表格中的数字。
答案：(1)星期四(2)星期六(3)星期二(4)1是蓝色(5)31是橙色

P11
这个游戏主要介绍"分配"的概念。先帮孩子读懂题目。可以借助方框来写出算法。
答案：(1)2 (2)2 余 1

P12
孩子估计数字的时候与实际的数字一般有误差。跟孩子讲明，估计数尽可能接近实际数字就可以。孩子不需要修改估计的数字，并且在写估计数字的时候不需要数点数。鼓励孩子看一下图片并且很快写出估计的数字。只有要看下一幅图片的时候才能具体数一下点数。

P13
这个游戏是帮助孩子了解0~5的数字关系。让孩子先算出各个算式，然后将算式结果与蜂巢相连。
答案：3+1=4, 4-1=3, 2+2=4, 1+4=5, 5-5=0, 4-3=1, 3+2=5, 5-3=2 (2)蜂巢号码数最大的是5, 最小的是0

P14
孩子需要衡量哪种活动花费的时间更短。跟他一起研究一下每种活动花费的时间长短。可以用两只空塑料瓶制作计时器。用沙子填满一个瓶子，用强力胶带把瓶子的顶部固定起来。在瓶子顶部打一个小洞，让沙子流出。让孩子用计时器来比较日常活动所需要的时间长短。

P15
这个跳房子游戏帮助孩子了解数字的顺序，并认识数字。
答案：(3) 1~9 (4) 11~20

P16
这个游戏让孩子了解另一个测量长度的非标准单位——脚的长度。鼓励孩子让双脚保持在一条直线上用脚跟顶脚尖的方法来进行测量。

P17
这个游戏是帮助孩子找到一组物品中哪些具有相同的属性。发现每一组物品和自己收集的物品中的相同之处是很重要的。画出自己收集的物品可以帮助孩子记录数据。
答案：(1)Jody 的物品都是洗漱用品(或用于清洁)
(2)Emma 的物品都是植物(或都来自于花园)
(3)Jaco 的物品都是体育用品(或用于游戏)

P18
这个游戏可以帮助孩子用"yesterday""today"和"tomorrow"将事件发生顺序进行排列。阅读句子时孩子在情境中选择最合适的词语。
答案：Yesterday, Today, Tomorrow

P19
与孩子一起读问题，并确保孩子已经理解了需要做的内容。或者也可以让孩子大声朗读出来。孩子需要按照题目要求来画出纽扣，并写出答案。让孩子先画出纽扣，再写出数字。将一个问题可视化呈现，是孩子学习过程中解决问题的一个有效方法。
答案：(1) 5, 2+3=5 (2) 4, 6-2=4 (3) Jody (4)9

P20
对于这个游戏而言，真正拿来一个盒子并在盒子前后摆放物品对孩子很重要。先让孩子了解物品放在盒子前后的位置概念。这个练习可以帮助孩子对物品的位置关系进行描述。

P21
这个游戏是给孩子更多的机会锻炼解决问题的能力。先和孩子一起读问题。然后让他在写出答案之前画出樱桃的总数。
答案：(1)5 (2)7-2=5 (3)8 (4)4+4=8

P22
这个游戏中，孩子可以2个2个地加，最终答出问题。通过这个练习可以帮助孩子更好地了解第一页中所说的"更多"的概念。
答案：(1)10 (2)8 (3)10 (4)10

P23
帮助孩子理解"一半"的概念最简单的方法就是将物品平均分给2个人。让孩子练习能将糖果均分给两个孩子。
答案：(1)5 (2)4 (3)3 (4)2

P24
这个游戏是在孩子认识数字1~10 的前提下，让孩子将图片和物品的数量相匹配。

P25
孩子能够计算图片中物品的数量，也可以看到每个数字的表示方法。让孩子数一下物品的数量并将答案写出来。这个游戏同时可以练习数字的写法。
答案：(1)9 个球, 6 个风筝, 7 个娃娃 (2)娃娃 (3)6 (4)more

P26
这个游戏可以帮助孩子了解直线边的物体和圆弧边的物体之间的区别。还可以同时帮助孩子了解同种物品大小的不同。举一些真实的例子：比如茶匙、汤匙和不同大小的碗。让孩子画完之后涂色，建议不同形状涂成不同的颜色。

P27
让孩子2个2个地数。如果他不太理解"向前"和"向后"的概念,可以让他想想年龄。比如,今年7岁,去年6岁,明年8岁。
答案:(1)8 (2)8 (3)7 (4)8 (5)6 (6)10

P28
让孩子理解"双倍"的概念就是一个数值加上相同的数值。如,4的双倍就是4+4。
答案:(草莓)4+4=8;(香蕉)3+3=6;(橙子)5+5=10;(柠檬)2+2=4

P29
这个游戏是让孩子理解完成一件事情相对需要的时间长短。孩子需要考虑完成一件事情所需要的时间。与孩子讨论一下,让他将所需时间较长的事情标记出来。

P30
这个游戏可以锻炼孩子根据给出的要求收集物品(数据),比如按要求的颜色收集物品。将收集的物品在图框中记录下来。让孩子数一数并记录下来收集了多少棕色和绿色物品。让他了解多或少的概念。

P31
了解数字顺序的向前数和向回数对于掌握加法和减法是非常重要的。让孩子用书中的"卷尺"来找到答案。
答案:16,20,17,20,34,17,30,23,14,30,10

P32
这个游戏帮助孩子掌握10以内加减法。第2、3问需要在已知的总数上做加减法。这个练习又可以复习加减法。
答案:(1)3+4=7,5+3=8,4+3=7,3+0=3,9-3=6,7-4,3+3=6,2+3=5,8-3=5,6-3=3 (2)4+2+2+2=10 (3)8-3-3=2

P33
这个游戏能给孩子测量的实际操作机会。请先准备好所需物品。先准备好一个约175mL的酸奶杯、金枪鱼或番茄酱这类罐头罐子。如果天气好,就让孩子在室外操作,这样也不怕水洒出来。

P34
这个游戏帮助孩子用序数词表示位置和排序。
答案:(1)Fatima (2)Mandla (3)第9名 (4)第6名 (5)第2名 (6)第10名 (7)第5名 (8)第8名

P35
这个游戏帮助孩子练习加法、计数、写算式以及数字的英文表达。在加法中,可以为孩子讲解相同结果的不同计算方法(如,4+5=9,5+4=9)。
答案:3+7=10,7+4=11,4+0=4,7+2=9

P36
在这里,您的孩子练习用多种硬币的币值相加可以等于另一个币值。
答案:(1)5分+5分+5分+5分=2角,1角+1角+1角=3角,5角+5角=1元,1元+1元+1元=3元 (2)7元

P37
孩子做游戏之前,先将8支蜡笔放在桌子上,让他估计一下数量,然后让他数一下具体数字。和他讨论一下估计数与实际数的差距。在练习中,保证孩子是先估计每个盒子中的蜡笔数,然后再实际数一下具体数。
答案:(1)8,6,10,8 (2)2 (3)1和4

P38
在这个游戏中,先让孩子描述一种规律,然后再用熟悉的形状和颜色创建一种她自己的花纹图案。第2题给的形状是有限制的,让孩子先画形状再填颜色。

P39
鼓励孩子先估计一下图中物品数量,然后再具体数数。让他迅速地看一下图片,然后就写下估计的数字。再让孩子对比一下实际数字和估计的数字。不要对估计的数字做出修改。
答案:7辆汽车,11扇窗户,7棵树,8个人,4幢建筑物

P40
让孩子找出数列的规律,并将缺少的数字填出。将训练数列的重点放在蛇形数列中。蛇形数列与传统的数列不同,需要对数字进行排列。
答案:(1)第1行:3,2;第2行:7,8;第3行:5,1 (2)1,2,4,5,8,9,10,12,13,14,15,17,18,19

P41
这个游戏能帮助孩子识别有直线或弧形边物品的形状。如果孩子认识一些形状(如棱柱体、三角形、长方形、圆形和球形),可以和孩子讨论这些形状的物品。
答案:(1)棒棒糖——球形,硬币——圆形,三角铁——三角形,球——球形,信封——长方形,录音机——棱柱体

P42
在这个游戏中,孩子可以通过多种方法练习加法。
答案:8,11,12,15

P43
这个游戏给孩子提供实际估测的经验。测量和记录的容量使用了比第33页更大的容器。因为在测量时容器需要装满水,最好让孩子在户外场地做游戏。

P44
重要的是,让孩子在日常生活中看到"对称"。可以先从日常生活物品中说明它们是如何对称的(如一个橙子从中间切开,一条裤子沿中线从左右折叠一下)。在最后一项练习中将关注点放在孩子理解对称概念上,而不是要画得有多完美。

P45
这个游戏可以帮助孩子练习2、5、10的倍数计算。孩子也能将手指分为2、5和十一组。当然要确保他知道手和手指之间计数的区别。
答案:(1)6,30 (2)8,40 (3)10,50

P46
这个游戏帮助孩子发现弧形边的物体可以滚动,而那些有直线边的物体可以滑动。在练习之前可以先找几本书、一块长方形的木块、几个盒子(棱柱)和一个球(圆形)让孩子体验一下滚动和滑动。
答案:串珠:I have curved edges. I roll. I am a sphere. 三角形:I have straight edges. I slide. I am a triangle. 礼品盒:I have straight edges. I slide. I am a prism. 圆形:I have curved edges. I roll. I am a circle.

P47
先在一个容器中放20个相同的小物品,让孩子估计数并数一下实际有多少个。然后让他看一下这些袋子的图片,估计每一个袋子中球的数量。让孩子自己数一下实际数量,然后对比一下估计的数量。
答案:(1)Lebo有15个弹珠,Jody有14个弹珠,Jaco有18个弹珠,Mandla有12个弹珠(2)Jaco(3)Mandla

P48
这个游戏帮助孩子了解"双倍"的概念。孩子要确定每个骨牌上有多少个点,并将正确的贴纸贴在骨牌上,然后将两个数加在一起。
答案:(2)8,16,18,14,12,2,10,4

P49
先确保孩子能数到20,并且在做练习之前可以数出20个物体。这个游戏是训练孩子对20以内数字的估测能力。
答案:(2)1号罐子中有8个小珠子,2号罐子中有12个小珠子,3号罐子中有17个小珠子,4号罐子中有18个(3)4号罐子中的珠子数量与Lebo的估计数最接近

P50
数字线的计算方法对帮助孩子理解数字概念很重要。在这个游戏的第二部分,鼓励孩子用数字线计算2的倍数。
答案:(1)13,12,8,15,18,6,12,9 (2)14只眼睛

P51
将散落的玩具进行比较和记录。这个游戏锻炼孩子将玩具分类,并记录和解读分类信息。
答案:(2)图片中共有2块拼图、6块积木、3个玩偶和4个毛绒玩具(3)积木最多(4)拼图最少

P52
这个游戏可以通过孩子感兴趣的方式教他如何使用钱币。你可以先拿出一些硬币,帮助孩子简单使用硬币购买东西。
答案:(1)a.爆米花 b.拐棍糖 c.棉花糖 d.拐棍糖和棒棒糖 e.云朵糖果 (2)可以有很多种选择组合 (3)4,2

P53
可以在这个游戏中锻炼孩子计数和创造数字模式的能力。还可以让孩子识别单数和双数。
答案:(1)30 (3)0,2,4,6,8,10 (4)5,8,11,14,17,20

P54
在这个100以内的数字模块中,可以帮助孩子了解数字的顺序。每行都是以"10"结束,以"1"打头。
答案:(1)3,5,12,14,17,20 (3)56,59,23,62,38 (4)100

P55
这个游戏可以帮助孩子进一步了解数字顺序,并且可以通过贴纸的方式介绍"位置"的概念。可以

让孩子先摆放一些厨房物品。给他5个不同的物品,要求他把物品放在一条直线上,并说明每一个物品的位置。鼓励孩子按从左到右的顺序来表达。

P56
这个游戏帮助孩子练习数字上加10和减10。

P57
让孩子先使用非标准测量单位测量的方法是很有必要的,比如先用手进行测量,然后再学着用尺子和其他测量仪器进行测量。在这个练习中,鼓励孩子按用手进行测量,不用强求测量结果特别准确。

P58
这个游戏帮助孩子继续练习"双倍"和"一半"的内容。让孩子先将所有题目算出,然后再涂色。
答案:(1)一半:20,5,6,3,15;双倍:40,10,8,18,16,14 (2)隐藏的数字:19 (3)40 (4)3

P59
这个游戏可以通过完成图画帮助孩子了解"对称"的概念。他可以通过画身体各部位体会身体的对称性,通过完成衣服上的图案,包括数字8,体会图案的对称性。

P60
这个游戏帮助孩子练习"相加"。孩子可以通过两个数字相加得到帽子中的数字。鼓励孩子用数字线进行计算。
答案:(1)12+8=20,15+4=19,16+4=20,15+6=21,6+8=14,16+9=25,11+1=12,11+9=20
(2)12+8=20,15+6=21,12−1=11,19−4=15,6+8=14,20−11=9,20−16=4,25−9=16

P61
天平不是每个家庭里都有。如果你的家里有,让孩子用这种非标准测量方式练习一下不同物品之间的平衡,如弹珠、硬币或者豆子。让孩子在用弹珠称重之前估计一下多少颗弹珠与物品重量相似。最后问一下孩子哪个物品最轻,哪个物品最重。
答案:(1)2,10,8,20 (2)最重的物品是鞋子 (3)最轻的物品是铅笔 (4)因为要使天平平衡,铅笔所用弹珠最少,鞋子所用弹珠最多

P62
这个游戏帮助孩子了解物品与盒子的空间关系。让孩子在画完一个物品之后,解释一下物品与盒子的空间关系。比如,橡皮在盒子后面。
答案:(6)棱柱体

P63
通过这个游戏让孩子对物品进行收集和分类训练。

P64 和 P65
这个游戏可以让你的孩子自己玩儿,也可以和更多的小朋友一起玩儿。只要保证孩子已经认识数字,并会计数,就可以让孩子愉快地游戏了。

P66
这个游戏帮助孩子解决一些问题(一些有余数,一些没有余数)。
答案:(1)每行种4棵雏菊,没有剩余
(2)每行种5棵玫瑰,1棵剩余
(3)每行种5棵郁金香,1棵剩余

P67
重复使用加法和减法是学习乘法和除法的第一步。鼓励孩子在练习中2个2个地数,5个5个地数和10个10个地数,并在图片中找答案。
答案:30个脚趾,4只袜子,10只手,8只手套,10只气球

P68
5个5个地数是一种有效的计算手指的方法。鼓励孩子使用数字线来进行5的计数。第二个练习的重点是5的加法数。练习中可以帮助孩子计数到20。
答案:(1)5 (2)10 (3)50 ,10+10+10+10+10=50
(4)15,16,10,4,6,20,1,3

P69
在补充完整每条项链之前,孩子要先观察,并找出其中的规律。
答案:0,5,10,15,20,25,30,35
100,99,98,97,96,95,94,93,92,91
80,70,60,50,40,30,20,10,0
35,34,33,32,31,30,29
12,14,16,18,20,22,24,26,28,30

P70
这个游戏可以帮助孩子找到日历中生日的时间。

P71
这个游戏中,先让孩子估计一下蜡烛的数量,然后再具体数一下。这次需要解决的数量比较多,教孩子先将10个分成1组,这样计算总数比较容易。孩子还可以用到5的倍数,一直计到100。帮助孩子用数字线5个5个地计数。
答案:(2)50 (3)20 (4)10,14,20

P72 和 P73
这个游戏是关于物品形状和空间描述的。可以用这些词语:在……后面、在……旁边、在……下面。你可以先拿一把椅子来为孩子做演示。比如坐在椅子上、站在椅子后面等。
答案:(2)门是长方形的。球在门的前面,是球形的。墙上的马赛克图案是三角形的。凳子上的洞是圆形的。大门上的洞是椭圆形的。门上的图案是长方形的。凳子上的盒子是棱柱体。
(3)Jody is behind the tree. Lebo is in the bush. Emma is under the bench. Ravi is behind the bush. Jaco is on the wall. Honey is next to Mandla.

P74
这个游戏帮助孩子使用金钱,并计算加减法。
答案:(1)2元、3元、4元、10元、20元 (2)兔子找回10元,球找回17元,熊找回0元,小汽车找回16元,棒棒糖找回18元 (3)a. six b. seven c. nine d. thirty

P75
这个游戏进一步强化孩子对形状的认识。可以在实际生活中找一些物品,看看他们是能滑动还是滚动。讨论一下物品是什么形状,比如,盒子是棱柱体。
答案:(1)滚动的物品:轮子、苹果、串珠 (2)滑动的物品:盒子、金字塔、尺子 (3)圆形和椭圆形是弧线的边(4)长方形、菱形、正方形和三角形是直线边

Certificate

证书

已通过了《奔跑吧,数学》2级闯关!

Smart~Kids

Make the smart choice for a brilliant future!

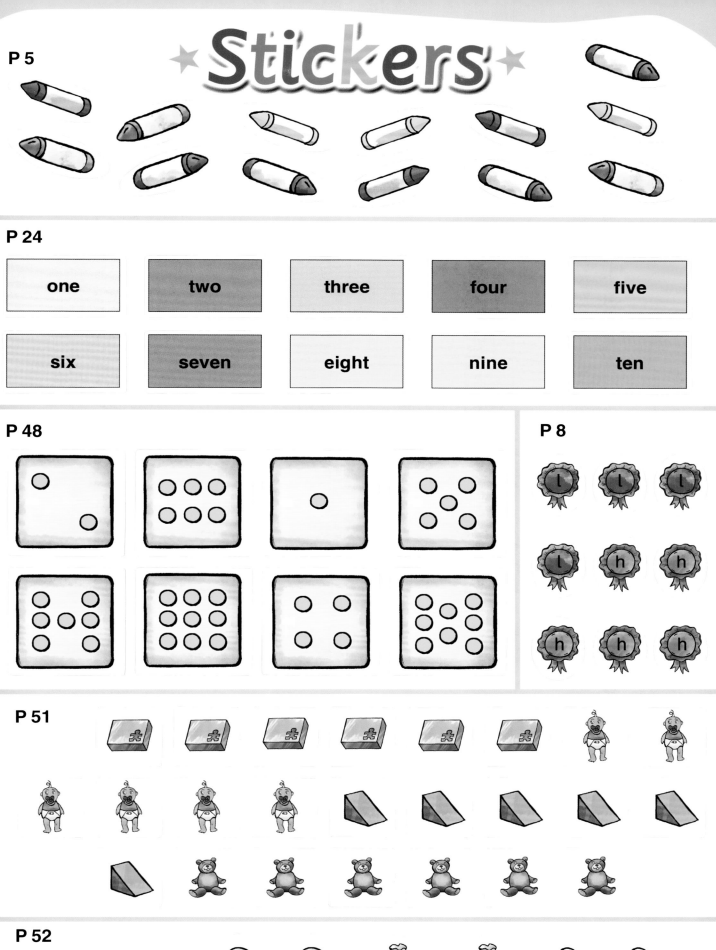

P 55

7th	12th	5th	17th
9th	14th	3rd	16th
18th	8th	2nd	20th
15th	4th	6th	11th

P 66

P 65